Constructing Chicago

Constructing Chicago

Daniel Bluestone

Yale University Press

New Haven and London

Publication of this book has been aided by a grant from
the Graham Foundation for Advanced Studies in the Fine Arts.

Set in Garamond Book type by Keystone Typesetting, Inc.,
Orwigsburg, Pennsylvania.
Printed in the United States of America by Arcata Graphics
Halliday, West Hanover, Massachusetts.

Library of Congress Cataloging-in-Publication Data

Bluestone, Daniel M.
Constructing Chicago / Daniel Bluestone.
p.   cm.
Includes bibliographical references (p.) and index.
ISBN 0-300-04848-3 (alk. paper)
1. City planning—Illinois—Chicago—History—19th century.
2. Urban beautification—Illinois—Chicago—History—19th
century. 3. Architects and patrons—Illinois—Chicago.
4. Architecture and society—Illinois—Chicago. 5. Chicago
(Ill.)—Buildings, structures, etc. I. Title.
NA9127.C4B48 1991
711'.4'097731109034—dc20   91-14675 CIP

The paper in this book meets the guidelines for permanence
and durability of the Committee on Production Guidelines for
Book Longevity of the Council on Library Resources.

10 9 8 7 6 5 4 3 2 1

Front endpapers:
Chicago in 1857. A section of the J. T. Palmatary/Braunhold
and Sonne panorama of Chicago. Chicago Historical Society.

Back endpapers:
Chicago business district panorama, 1898. Library of Congress.

For Joanne Baxter Bluestone

and to the memory of Max Bluestone

Contents

Acknowledgments

This history, of course, has its own history. This book started as a dissertation. Neil Harris's work as a historian, teacher, and adviser inspired my approach to Chicago's architecture, landscape, and culture. At Chicago, I learned a tremendous amount from my friend Ken Cmiel as we studied along parallel paths; "his people" attempted to construct a language of democratic eloquence; "my people" attempted to construct a city of cultivated sensibility. The fact that "our people" shared common ground in the nineteenth century has nicely complicated and expanded the range of this book.

Many people have my special thanks. Kathleen Neils Conzen, James Gilbert, and Roy Rosenzweig read early versions of my manuscript and offered their own special perspectives as historians of nineteenth-century urban America. Donald Olsen burst the bounds of the "anonymous reader" genre and responded to my penultimate draft with a generosity and critical engagement that made the final work on this book both a challenge and a pleasure. Areta Pawlynsky drew on her Chicago youth, her Columbia training, and her abundant good will to create several drawings for the text. Philip Krone generously supported this project with his uncanny insights concerning the people and buildings of Chicago. The staffs and resources of the Chicago Historical Society, the Library of Congress, the Avery Architectural and Fine Arts Library at Columbia University, the Mrs. Giles Whiting Foundation, the Smithsonian's First Ladies Fellowship program, and the Graham Foundation provided important support for my research on Chicago. Judy Metro and Lorraine Alexson, my editors at Yale University Press, have been a pleasure to work with. In moving the manuscript from the desk and the disk to the bookshelf they have combined enthusiasm with precision, and humor with thoughtfulness.

Barbara Clark Smith came to nineteenth-century Chicago from eighteenth-century Massachusetts. Her formidable insights and skills as a historian have alternately challenged and enriched my own construction of Chicago history. She has read and discussed and debated all that is here. Her special ability, honed as a Smithsonian curator, to insist that complex historical interpretations be presented in straightforward and pithy prose has prompted a level of clarity for which I and my readers are deeply indebted. Barbara has, at the same time, shared equally in the raising of our children, Hattie and Hal, who have given us no small measure of joy in our own twentieth-century history.

Constructing Chicago

Introduction

The sheer scale of commerce, the gritty industrialism, the materialism and greed of nineteenth-century Chicago have consistently dominated views of that city. Carl Sandburg's "Chicago," for example, reverberated with material developments but little else: "Hog Butcher for the World, / Tool Maker, Stacker of Wheat, / Player with Railroads and the Nation's Freight Handler / . . . / Stormy, husky, brawling, / City of the Big Shoulders." Sandburg described a city that proudly reveled in the vitality of hard work honestly done: "Bragging and laughing that under his wrist is the pulse, and under his ribs / the heart of the people, / Laughing! / Laughing the stormy, husky, brawling laughter of Youth, half-naked, / sweating, proud . . ."[1] Whether they have lamented or celebrated the nature of the city, few have disputed Sandburg's selective emphasis. The most enthusiastic urban boosters, the most critical urban reformers, the most engaging novelists, poets, and historians have grappled with Chicago's staggering commercial and industrial expansion, convinced that the largely untempered forces of capital have defined the city. In this view, Chicago represented much that was quintessential about the American nation: It was raw, immature, and ambitious, lacking culture or history but seeking wealth and material growth, itself a product of the frontier far more than of Europe and the civilized east. Here, in the nation's nineteenth-century heartland, was a city unresistant to Mammon; here in Chicago was "the land of dollars."[2]

This idea has profoundly shaped interpretations of Chicago's nineteenth-century architecture and landscape. Surveying the course of American urbanization, Lewis Mumford wondered, "Did ever so many elements of disintegration come together at one time and place before?" In the end, he decided: "That a city had any other purpose than to attract trade, to increase land values and to grow is something that, if it uneasily entered the mind of an occasional Whitman, never exercised any hold upon the minds of the majority of our countrymen. For them, the place where the great city stands *is* the place of stretched wharves, and markets, and ships bringing goods from the ends of the earth; that and nothing else."[3] What was generally true of American cities, moreover, was particularly true of the great entrepôt of the Midwest. In Chicago especially, city building proceeded largely uninformed by aspirations to beauty, art, or repose.[4]

Aspects of architecture and urbanism devoid of aesthetic refinement seemed, in this view, to be particularly *Chicagoan*. Not surprisingly, the skyscraper takes a central role in this account of the city's architectural history. Soaring to unprecedented heights, the tall office building was undeniably a tool for private money making. Its self-assertiveness, its bold incorporation of modern technologies, its rapid monopolizing of limited downtown space—all marked it as an expression of the quest for profit. Equally important, critics have emphasized the Chicago skyscraper's contributions to an architectural style that, blossoming in the twentieth century, cut ties with the aesthetics of the past, eschewed ornament and artifice, and frankly acknowledged underlying structure. Scholars have thus drawn a narrative line that links Chicago skyscraper architects such as Louis Sullivan directly to twentieth-century modernists such as Mies van der Rohe.[5] Little touched by history, Chicago's skyscrapers were necessarily modern. And while these structures most eloquently expressed Chicago, other aspects of the metropolis—whether urban parks, churches, or cultural and civic buildings—have seemed secondary by contrast, for although Chicagoans undeniably lavished time, money, and design concern on these creations, in doing so they were often participating in wider national and international movements. Arguably, then, these aspects of the city were derivative, inexpressive of Chicago's exceptional character. Skyscrapers alone were emblematic of the city, tapping its indigenous energy. According to this formulation, the fact that Chicagoans celebrated an aesthetic expressed in skyscrapers argues the inauthenticity and the tenuous hold of more traditional notions of beauty and art.[6]

This book presents a different interpretation of the culture and architecture of nineteenth-century Chicago. It disputes the view of Chicago as singularly or solely preoccupied with Mammon, for although it does not deny the real force of money making in shaping Chicago, it considers profit taking as a cultural practice, framed by a culture that channeled it and gave it meaning. Most important, perhaps, it contests Mumford's characterization of nineteenth-century urban dwellers as unreflectively consumed with material growth. Chicago's city builders viewed architecture as an important means of expressing and effecting social and cultural ideals. Often self-consciously, they set out to build not just a place to make a profit but a place to inhabit and hence an arena for particular ways of living. As a result, nearly from its founding, Chicago was expressive of more than the quest for profit; it expressed a series of learned and shared aspirations for urban life. Notions of aesthetics, civility, and moral order deeply influenced the city that rose so dramatically on the shores of Lake Michigan. There is no understanding nineteenth-century Chicago's architecture and landscape without taking account of them.

This, then, is a study of how culture made itself manifest in Chicago's nineteenth-century cityscape. It traces the ways that cultural ideals expressed themselves in an urban aesthetic: a sense of appropriateness regarding the site, use, form, style, and expense for buildings and spaces within the city. I present an alternative narrative of the city's architectural past, reordering various building forms and reassessing their significance. In this view, the nineteenth-century skyscraper fits more comfortably with its own contemporaries than with twentieth-century successors; nineteenth-century skyscrapers appear less exceptional, more grounded in the past. It follows that parks, churches, and civic and cultural institutions reflected nineteenth-century Chicago as truly as skyscrapers did—and even, I will argue, in much the same way. Each of these forms, unquestionably shaped by the material standards and money-making desires of capitalist Chicago, also felt the impact of Chicagoans' desire to express and maintain commitments beyond the market.

My concern in tracing the links between city building and culture marks this study as a particular approach to architectural history. I move away from specifically architectural questions about the relation between one building design and another or between a building design and an architect's formal models, precedents, influences, and innovations. Studies that focus on these questions have provided important clues as to why a particular building has taken a particular form. Such questions, however, stop short of explaining the cultural question of why a building is built at all. I am interested in understanding architectural expression as it was originally conceived by the city builders within their social context. Although architecture may be distinguished from vernacular building on the basis of its pretensions to academic "art," it rarely represents a capricious or isolated act of individual creative genius. Public buildings—and the buildings with which private parties, such as corporations, voluntary organizations, and religious congregations, present a public face—are most particularly institutional and social expressions as well as individual ones. Chicago's nineteenth-century architects were often creative and sometimes inspired. In the pages that follow I do attend to their ideas about their work, but my focus is equally on the expectations and ideals of the entrepreneurs, church officials, and civic leaders who were architects' clients. For the most part, I treat architecture less as an individual product than as a social one.[7]

Of course, the architectural landscape of cities such as Chicago was not related to urban culture in a simple or reflexive way. In part this is to say that there was nothing simple about *urban culture* in the nineteenth century.[8] Chicago began as a swampy frontier outpost with barely fifty inhabitants in 1830; within decades it was a teeming, varied city. The city doubled and redoubled its population at an astounding rate: nearly 30,000 in 1850, nearly 300,000 only twenty years later. Railroad tracks, stockyards, lumberyards, and factories appeared at a staggering rate, and measures of industrial output multiplied accordingly. Irish, German, and Eastern European immigrants crowded into the city in turn. By 1890, Chicago was America's second largest city, with a population of over 1 million. It was the home of many cultures and the site of bitter class conflict.[9]

Not surprisingly, this nineteenth-century city expressed the aspirations of some Chicago residents more than others. Those who most powerfully shaped the urban landscape certainly formed an elite, a minority that wielded an influence far disproportionate to its numbers. From the founding of the city, Chicago's city builders were men of business—speculators in real estate, merchants in lumber, wheat, beef and pork, builders of railroads, manufacturers of iron and steel, as well as successful members of the professions.[10] (Architects might thus rank as members of this class, and they almost invariably worked for those who did.) At times, indeed, the city builders might easily be counted and named: very few Chicagoans, after all, could command the resources to raise a downtown skyscraper.[11] Chicago as constructed was not merely "social," then; it was the construction of particular members of a particular social class.

Although their wealth arose in the world of capital, in cultural terms city builders were knit together by a cultivated sensibility and aspirations to a style of life that was in some ways aristocratic, in some ways bourgeois. That knitting together emerges as one of the themes of this study, for over the course of the nineteenth century, elite Chicagoans elaborated a sense of themselves as the sort of people who would willingly take responsibility for the city, broadly defined. If most members of the *urban establishment* moved to Chicago to make fortunes, they soon began assessing the city in a new way. What were the bases of their engagement with their city? Contrary to Mumford, Chicago's city builders explicitly expressed unease about the reputation of Chicago as the heart of Mammon—and, in fact, about the reality of their class's commitment to profit making in a capitalist city. They projected that cultural problem markedly into the work of city building. To express values transcending the material, moreover, they generally adopted standards of beauty and civility derived from European and historical sources. Far from glorying in their status on the frontier or celebrating a rough-hewn Americanism, elite Chicagoans continually worried about the impression their city made on visitors from more cosmopolitan places.

Equally important, to lay claim to civility and culture, Chicagoans progressively segmented their city, creating separate and insulated arenas for different activities and aspects of life. Wealthy Chicagoans, like their counterparts in other cities, lived in an increasingly fragmented world in which work and leisure, intellect and emotion, male and female were held separate and counterpoised.[12] The city they imagined was correspondingly fragmented. They built some parks and churches, for example, specifically to secure a domestic and residential sphere apart from the world of work. They sought physical expression of identities determined not in the marketplace but in other, supposedly purer, urban realms.

Recognizing this point, historians have often characterized parks and residential enclaves as "retreats" from the nineteenth-century city. This formulation seems inadequate on several counts. As they created these arenas, nineteenth-century Americans were not moving backward toward eighteenth-century patterns of urbanism nor toward existing forms of agrarian life—both of which were marked by a customary physical coincidence of workplace and residence, by the interpenetration of production and reproduction, and by a cultural sense of the compatibility of these. Similarly, far from moving closer to nature, nineteenth-century urbanites were rather creating new and artificial forms of it—the park, say, or the family—and making innovative claims about it. Finally, there was little in the creation of the segmented city to suggest a profound streak of *antiurbanism*. Instead, Chicagoans constructed realms sheltered from the influence of Mammon or, more accurately, realms from which the force of Mammon was veiled or symbolically excluded and where, therefore, they might harbor values inappropriate or inimical to the commercial downtown. For these reasons it is more useful to consider parks, residential enclaves, and suburbs as fully parts of the modern city rather than as modifications of or reactions against it.

This view challenges nineteenth-century claims that residential, cultural, and religious domains actually stood *apart* from commerce. Such categories as work and leisure, public and private, were being formed in the course of the century. The assignment of different building types to these different domains of life was a historical process that we can observe under way in Chicago. Equally, Chicago's city builders produced a gendered cityscape; they formed new ideas about the appropriateness of middle-class men and women's presence and activity within different parts of the city.[13] It is notable, therefore, that the city builders themselves were not only of a specific class but, by and large, of a specific gender. As they assumed responsibility for culture, beauty, and morality in the city, these men expressed their own ideals of gender. It is possible, indeed, to read the landscapes they created as efforts to navigate between commitments to private and public lives.

Finally, although segmentation was a real and pervasive strategy affecting the nineteenth-century cityscape, nonetheless the impulse was not unalloyed. In some parks, exceptional churches, and commercial buildings such as skyscrapers, the city's elite often expressed a sense of the benign possibilities of urban America, which sometimes included the potential benefits of social heterogeneity. The elite explored the possibility that downtown commerce might support, rather than menace, refined civilization. In addition, even the harshest critic of commerce might recognize limitations in the spheres symbolically denied to Mammon. Constructing the city in this way did not always allay the anxiety of city residents about commercial development or mediate the conflicts created by it. It follows that if at times the Chicago landscape reflected elite residents' primarily intraclass preoccupations, at other moments or in other venues city builders' concern over interclass relations figured as well.

Several factors influenced the city elite's preoccupation with themselves or engagement with others. After the Civil War, what had been a fragmented group of prospering settlers achieved relative cohesion among themselves and, with it, a sense of stability that encouraged responsibility for instilling ideals of cultivation outside their own ranks.[14] As Eric Hobsbawm has recently noted, the boundaries of the middle class shifted in the course of the nineteenth century; by the late 1800s, new groups of managers, clerks, and other white-collar workers aspired to the respectability of bourgeois status.[15] These were rarely the most articulate or powerful proponents of city building strategies, but, as the chapter on skyscrapers maintains, their presence made itself felt among those who were. Equally vital, the presence of Chicagoans unquestionably outside the ranks of the middle class and relatively immune to middle-class aspirations increasingly influenced city building strategies toward the end of the nineteenth century. Evidently, too, the freedom of the elite to shape its city varied from one urban form to another: builders of private commercial structures had to accommodate few besides potential building tenants. If they aspired to art, they might consider the opinions of journalists and other arbiters who discussed Chicago's architecture in the contemporary press, but this latter group surely reached print at least partly by virtue of its limited independence from the city's elite. Similarly, vestrymen, wardens, and trustees of

Chicago's Protestant congregations needed to consult the ideals of few outside their own social circles. By contrast, to build extensive park systems or civic centers often meant securing the consent of voters. When voters defeated a park system referendum in 1867, the defeat indicated the tenuous hold of wealthy Chicago residents on local politics. In the face of the rising political power of working-class Chicagoans, the elite turned to private reform organizations and more concerted efforts to extend the influence of its own cultural ideals.[16] To varying degrees, then, Chicago's city builders made efforts to enlist their social inferiors in their own social vision and at least to assign them a place within it.

I have organized the chapters that follow thematically, focusing in turn on parks, churches, skyscrapers, and civic and cultural buildings, along with the public (and occasionally private) discussions that took place about each of these forms. At the same time, there is a rough chronological sequence. Although the city's park movement came to fruition after midcentury, its roots lie in the early years of settlement when the city designated the lakefront as a public ground and when city gentlemen began to pursue ornamental horticulture on their own estates. And although Chicagoans first considered the civic landscape when they built a city hall in the 1850s, their most ambitious and coherent efforts to secure the civic presence centered on plans for the lakefront created in the late nineteenth and early twentieth century. Taken together, then, the chapters move from the city's founding in 1830 to the 1909 Plan of Chicago. If a neat symmetry seems apparent in thus beginning and ending on the lakefront, it might suggest the continuing significance of that site to public Chicago. Not least, since the Chicago River rather than the lake provided the city's harbor, the lake might be valued for its natural beauty and its immunity from commercial and industrial disfigurement. Nevertheless, this neatness is essentially artificial; it should not mask substantial changes amid continuity. Many developments in Chicago's park, religious, commercial and civic landscapes responded to the city's continual and remarkable growth.

In chapter 1, I look at the antebellum development of Chicago's public grounds and the movement for an extensive park system that came to fruition in the late 1860s. I examine numerous ways in which those movements were influenced by the promise of economic profit for holders of city real estate but suggest that a conviction of the insufficiency of economic striving spurred parks promoters as well. In newspapers and other public arenas, some Chicagoans explored the tremendous symbolic value of land that was explicitly withheld from commercial improvement; they elaborated their thinking about the meaning of *nature*, its associations, and its alleged effects on individual character and social relations. They forged a view of urban living in which selected and controlled aspects of nature played a crucial role. Tracing the development of Chicago's social conception of ornamental horticulture, I trace park promoters' growing self-consciousness as members of a city building class. In the park movement, some Chicagoans took on responsibility for embodying their ideals in a built environment. In the course of the century, moreover, they gave increasing thought to the ways that parks might mediate their relation to other classes, paradoxically promising both social separation and social unity.

I take up in chapter 2 the specific designs proposed for parks and boulevards within the city. I consider park designers, such as Frederick Law Olmsted, Calvert Vaux, William Le Baron Jenney, focusing on their practices, writings, and efforts to tutor Chicagoans in a preferred aesthetic. Looking at parks proposed and built, I suggest how these designers molded the landscape in hopes of affecting individual consciousness and social relations. Design elements, such as lakes and greenswards, and competing styles, such as the picturesque and the beautiful, all took on value for their supposed impact on human sensibility. The choices of designers were thus related to the wider social purposes of urban parks and boulevards. Indeed, the varied constituencies that supported and used parks made their own priorities for city parks felt in the decisions of designers. In effect, the resulting park and boulevard systems embodied rather divergent social visions.

In chapter 3 I explore the dramatic reordering of the urban religious landscape that began near midcentury in Chicago. Church builders participated in an international stylistic movement, the Gothic Revival, that proclaimed a connection with history. At the same time, they abandoned the forms of central religious monumentality familiar from childhoods spent in eastern towns. They devised novel patterns of church location. Underlying both of these developments, I suggest, was a conviction of the importance of distinguishing realms of spirit and Mammon in palpable landscape terms. Along with some urban parks, relocated churches contributed to the development of domestic residential enclaves removed from commerce. Whether churches were public (hence proper to the downtown) or private (hence appropriately sited in new residential areas), represented a substantial issue, even to elite Protestant congregations whose denominationalism did not promote social inclusiveness. Notably, during the 1870s and 1880s a few leading Protestant heretics challenged the assumptions

that created a decentralized and distinctly noncommercial religious landscape. As they attacked what one person termed the "limestone Christianity" evident on "the avenues," they underscored both the contradictions and the commitments embodied in the reconfigured religious landscape.

Movements for public parks and the creation of an impressive religious landscape both form a background for a reinterpretation of Chicago skyscrapers. In chapter 4 I consider skyscrapers in the context in which they actually arose—that is, in the midst of other efforts to create a city ennobled by commitments to culture and beauty. Reviewing the first generation of skyscrapers in Chicago, built between 1880 and 1895, I argue that, even in these designs for the commercial world, some Chicagoans preferred artistic embellishment and an abiding involvement with ideals of cultivation of the self and of a broader society. By highlighting aspects of skyscraper design most noted by contemporaries—monumental entrances, embellished lobbies and artistic elevators, spacious light courts, and overall height—I indicate how architects sought to mask the commercial basis of tall buildings. Skyscrapers were workplaces; they were also icons of a purified urban commerce. For these reasons, nineteenth-century Chicagoans simultaneously lamented and promoted the tendency of observers to equate the skyscraper with the city.

In chapters 5 and 6 I survey the rise of civic and cultural buildings in Chicago. Chapter 5 outlines early efforts to form distinctive and symbolic civic buildings. In the antebellum period, commentators imagined the city hall and county courthouse as expressions of a civilized and prosperous community, one capable of turning aside from commerce to build its grandest buildings for public accommodation. Postbellum cultural institutions displayed increasing attentiveness to site possibilities outside the downtown area and to the importance of a mutual relation among buildings. In chapter 6 I analyze the culmination of thinking about these elements in the various plans for lakefront civic centers proposed in the 1890s and 1900s. In those plans designers sought to coordinate the development of disparate civic buildings on massive public plazas. I argue that their plans responded to the city's most prominent commercial forms—factories, railyards, and above all else, the downtown skyscraper.[17]

In an epilogue I consider Chicago's place in the wider reaches of nineteenth-century American urbanism and explore the significance of Chicagoans' reservations about Mammon and their commitments to beauty and culture. I suggest how we might understand a Chicago that is not the city described by Sandburg and Mumford.

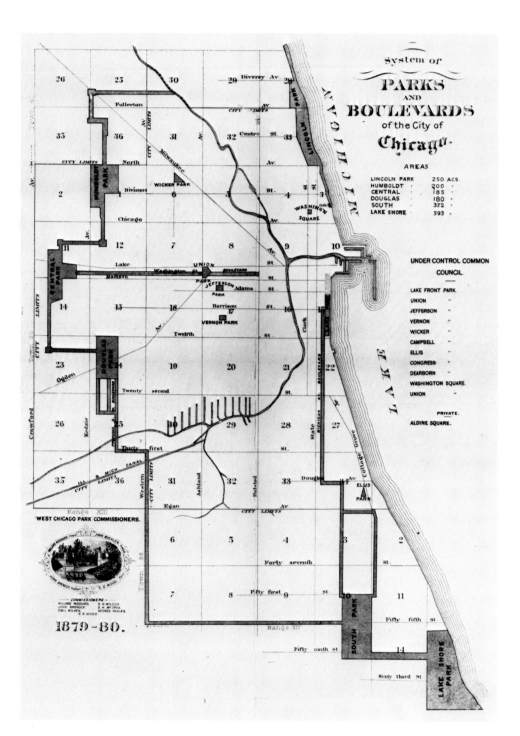

Chicago Boulevard and Park system, 1879–80. Map published
in West Chicago Park Commission annual report, 1880. Library
of Congress.

1

"A Cordon of Verdure" Gardens, Parks, and Cultivation, 1830–1869

The Illinois legislature aimed to give the burgeoning city of Chicago the "finest system of public parks in the world" when it passed the Chicago park bills in 1869. Outlining parks on all sides of the city, linked by miles of landscaped boulevards, the legislation impressively expanded Chicago's park area from under one hundred acres to over eighteen hundred acres. With this single gesture, the government created a new, expansive public realm within the city, reshaping neighborhood patterns and providing new boundaries for Chicago's future growth. What accounted for this dramatic reconfiguration of the city? Why did park promoters—a class of men accustomed to turning a profit through Chicago real estate—remove so much land from the purview of private ownership and exchange? Why did they exempt so many acres from commercial and residential development? Why, in short, did a grand system of public parks seem such an important part of the emerging urban landscape?

The creation of Chicago's parks had direct economic implications. Some people turned a profit on parks, and pecuniary calculations hedged about the entire park movement. Yet in Chicago park promoters were moved as well by particular ideals of *cultivation* that informed a changing nineteenth-century concept of urban life and urban form. The ideal of cultivation encompassed the physical designs, the social aspirations, and the economic vision of the park movement. Cultivation also expressed something of the complexity of the contemporary view of parks. For landscape design and planning, its associations were in many ways natural and rural. However, in relation to social aspiration, the term suggested urbane, cosmopolitan refinement. Many urban residents viewed park development as the act of a *civilized* community, a community capable of turning away from the pursuits of commerce to create a public realm for refined leisure. Urban real estate interests looked to natural landscape design for the cultivation of property values and the growth of the city. Simply put, idealizations of urban as well as rural life informed the notion of cultivation that guided park development. By the end of the century, numerous public acres of landscaped meadows, wooded areas, promenade grounds, carriage drives, and boulevards testified to the power of this ideal in Chicago.

Ideals of cultivation are central to the Chicago park movement that transformed the city in the late 1860s and beyond. They also shaped the city's antebellum landscape development, receiving formative local expression in the city's earlier private gardens and public grounds. Although the park system totally eclipsed efforts of the 1830s and 1840s at ornamental horticulture and public recreation, there were strong lines of continuity. Early devotees of private horticulture and the established tradition of public promenading, for example, helped shape the later public parks. Rather than suggesting a retreat from noisome urban conditions, these landscape improvements reflected a broad advance toward modern urbanism. In many cases they represented an effort to adopt the forms of leading American and European cities. Equally important, the cultivated city landscape bespoke aspirations toward refined individual character and a harmonious urban society.

Private Gardens

Long before Chicago could boast an extensive park system, private residents of the city pursued the cultivation of nature. William B. Ogden, often called the "Father of Chicago," played a key role in fostering private garden horticulture and in establishing the city's early public grounds. His views of Chicago's commercial and cultural life also anticipated the terms of later debates over the park system. Ogden, the son of a lumberman and merchant, was born in 1805 in Delaware County, New York. In 1834 he was elected to the state legislature, where he promoted the construction of the New York and Erie Railroad Company and other internal improvements. He was one of many nineteenth-century Americans who envisioned national prosperity and greatness, to be achieved through technology and enterprise. His appreciation of the importance of internal improvements in commercial development framed his view of Chicago, where he moved in 1835 as the agent for his brother-in-law's substantial real estate investments. In Chicago, with a population of only four thousand but with boundless promise, investors who purchased land and could weather periodic economic crises stood to make a fortune.[1]

From the 1830s through the 1870s, Ogden invested large sums of money in Chicago for himself and many eastern capitalists. In part to ensure profits from land, Ogden undertook a set of diversified investments. He led the promotion of, and served as a contractor for, the canal connecting Lake Michigan at Chicago with its immediate hinterland and the Mississippi River valley. Later he promoted Chicago's first railroad. He was also a banker, a newspaper publisher, and founded a lumber company. By all accounts, he was one of Chicago's wealthiest residents. He endowed the Rush Medical College, the Chicago Historical Society, the Academy of Sciences, several local churches, and the first University of Chicago. For decades Ogden stood as a leader in elite social and economic circles.[2]

The control and exchange of city land secured his position. For Ogden and others like him, Chicago's land took on value as it was removed from nature and possessed by enterprising men—as it was artfully modified, "improved," and located in a social frame. It followed that Ogden celebrated development—the art of improving on nature.

Yet although Ogden and other members of Chicago's elite prided themselves on the city's growth, they nonetheless expressed concern over the social character formed in a city most readily compared to a depot or entrepôt, where Mammon loomed so large. In 1837, for example, he declared that business speculation "tended to destroy the discriminating distinctions . . . and . . . integrity of character," so that "deceptive treachery and bad faith have for a time been awarded with the greater success." As a result, "a precious set of scoundrels have . . . positions in society which they are in no way entitled to. Others who have heretofore been counted honest men, too are now daily proving themselves to have been very corrupt."[3] Under the weight of speculative excitement and the economic crisis of the late 1830s, Ogden found that Western man's integrity, honor, character, and civic statesmanship were "very much like the paddy flea[:] 'put your finger on him and he's not there.'"[4] He stood at the very center of real estate speculation in Chicago, yet the results of speculation left him uneasy.

Ogden was one of several men of his class who felt that an antidote for the ills brought on by speculation lay in a different sort of manipulation of the city's land. Ornamental horticulture, to Ogden, represented a means of redressing the morally corrosive effects of business pursuits. Ogden fashioned his personal style of living accordingly. His Chicago garden embellished a rather grand house that occupied a four-acre lot bounded by Rush, Ontario, Cass, and Erie streets on the north side of the Chicago River. John Van Osdel, a professional architect and builder whom Ogden had convinced to move to Chicago, designed Ogden's Greek Revival-style house with a large piazza and a projecting pediment supported by columns. In developing his residence lot, Ogden showed a "perfect contempt of Yankee vandalism" by preserving the site's maples, elms, poplars, and ash trees—showing apparently unusual appreciation for existing landscape elements. In the late 1830s, Ogden transplanted the wild and ornamental shrubbery he had found in the Calumet Lake region, just south of Chicago, to his growing garden. He brought Carolina roses, dogwoods, red osier, kinnikinnink, Virginia creepers, bittersweet, lilies, and other wild vines. Graveled footpaths wound through Ogden's garden and grounds. He also maintained an extensive glass conservatory for growing flowers and fruits. In the 1840s he hired a European-trained gardener; in 1860 a Austrian-born gardener lived in the house along with Ogden's personal secretary, his coachman, five domestic servants, and business associate Edwin Sheldon and Sheldon's family.[5] Throughout these decades, Ogden devoted considerable time and expense to the landscape surrounding his residence (fig. 1).

In the first instance, these efforts reflected Ogden's character as (in a friend's term) a "real lover of nature."[6] His favorite poet was William Cullen Bryant, whose poems evoked scenes of a natural landscape that reminded Ogden of his childhood in Delaware County. This nostalgia combined with a heavily romanticized view of the ennobling effects of being close to nature. Thus, in 1839 Ogden wrote a friend in Sault Ste. Marie that he longed to visit her "quiet house & enjoy nature with you, pure and uncontaminated by the strife for gain or place which so degrade this busy world—I want to know how you breathe and think when freed from the contests of ambition, the lists of politics & tricks of the law . . . the temptations & snares of the crowd, the parade of wealth & fripping of fashion." Ogden declared that he would "gladly exchange the topics of the day in the bustle of society for the purer conceptions of uncontaminated minds . . . disconnected from every surrounding circumstance that is calculated to divert the thoughts from the purer channels of virtue, philosophy & religion."[7]

He prized contact with the natural landscape, then, for its supposed benefits to individual character and state of mind. Short of abandoning city society, cultivating a garden represented a means of softening the impression of urban life. By implication, hours spent tending one's garden or peacefully contemplating its beauties might counteract the effects of the daily pursuit of Mammon. Horticulture might also smooth and adorn social relations. Ogden entertained many friends and business associates at his estate; he was noted for carrying fresh flowers and fruits to his friends. [8] Ogden used his garden to ease and ennoble relations with his social peers.

Ogden's gardening, his love of nature, and his longing for the scenes of his childhood clearly acted as a balm for his distaste for Western Mammonism. His flights of nature fancy suggested the manner in which scenes of nature, quite removed from agrarian tradition, could be

invested with a distinct immunity from utilitarian and materialistic considerations. Untouched by social and material debasement, scenes of nature appeared pure. When urban residents incorporated elements of nature into the city, this vision of their purity and simplicity proved particularly attractive.

Yet, despite his declarations, Ogden stayed for many years in the "bustle of society" as a Chicago booster and speculator. He stayed longer than he had any financial need to stay, preferring a city garden to a country retreat. Indeed, his carefully tended grounds reflected less an idealization of nature than an adherence to an urbane ideal. It was not merely that Ogden altered the appearance of his grounds substantially—in fact, the landscape testified to Ogden's commitment to active improvement rather than contemplative serenity—the expression of personal character he sought through horticulture was neither raw nor natural. On the contrary, he pursued a highly refined and polished self that could only be the product of a thorough education and exposure to civil society. To desire cultivation was to seek a far different relation to nature than what Ogden touted in his writings.

Ogden embraced sophisticated high culture rather than rusticity. He appreciated music and poetry; he owned a large collection of paintings and engravings, including works by Church, Cropsey, Durand, Healy, Kensett, Powers, and Rossiter. He maintained an extensive library and a collection of statuary purchased in Europe. Visitors to Ogden's home found "generous and liberal hospitality . . . refinement, broad intelligence, kind courtesy." Ogden often entertained the most important visitors who traveled to Chicago. His guests included Fredrika Bremer, Charles Butler, Margaret Fuller, William Gilpin, Harriet Martineau, Joel Poinsett, Samuel Tilden, Martin Van Buren, and Daniel Webster. The life he lived was worldly, sophisticated, and urbane.[9]

In that life, horticulture played a double role, that of both a social and solitary pursuit. Ogden shared his love of landscape gardening with many others among Chicago's elite. Jonathan Y. Scammon, wealthy lawyer, banker, railroad promoter, and insurance company president, actively participated in the postbellum movement that established Chicago's park system and maintained one of the city's finest private gardens. In 1860 a gardener, L. P. Ostman, together with his wife and daughter lived in Scammon's Michigan Avenue household.[10] One of the city's earliest white settlers and a wealthy real estate developer, John H. Kinzie had a few acres of shade and ornamental and fruit trees around his residence,

1. Panoramic view of north-side houses and gardens. Detail of the J. T. Palmatary/Braunhold and Sonne, 1857 panorama of Chicago. The house with a cupola and garden in the center foreground is William Butler Ogden's. It is adjacent to St. James Church and is surrounded by the neighboring houses of Isaac Arnold, H. H. Magee, Walter L. Newberry, and Mark Skinner. At upper right is Holy Name Church. Chicago Historical Society, ICHi 05662.

which stood in the same wealthy residential enclave as Ogden's house. George W. Snow, a lumber merchant credited with the invention of light-weight balloon-frame construction, kept a garden as well as an extensive greenhouse that included ninety varieties of camellias and twenty-five varieties of cactus. Dr. William B. Egan, a medical doctor, real estate dealer, and a founder of the Chicago Lyceum who was also considered a classical scholar and an authority on Shakespeare, cultivated strawberries, vegetables, and flowers and maintained a large nursery garden with a variety of evergreens, which he transplanted to the grounds of residences along the lake shore. Merchant Henry G. Hubbard had the reputation of being Chicago's most successful transplanter of forest trees. Justin Butterfield, a lawyer who drew up the 1842 canal bill to ensure the completion of the Illinois Canal, possessed a "very tastefully laid out" garden with a greater variety and profusion of flowers than any other Chicago garden.[11]

In 1847 these elite gardening interests organized the Chicago Horticultural Society. John H. Kinzie assumed the presidency; Ogden, Egan, Samuel Brooks, a florist, and Ezra Collins, a manufacturer of boots and shoes, served as vice-presidents. The society established committees covering different fields of botany, disseminated useful gardening information, and held competitions and exhibitions.[12] In their society and gardens, Chicago horticulturists sought to foster a refined appreciation of their art. Acknowledging that observers judged the character and morals of people by the "external marks and signs that are so sure indications of internal states," Chicago's elite congratulated itself on the extensive private gardens that "mirror[ed] . . . a highly cultivated mind."[13]

In prizing such accomplishments and the way of life surrounding them, Ogden and others of his class were influenced by an ideal pursued by many aristocratic Americans of the preceding century. Early in the 1700s, elite Americans had begun to follow the English gentry in embracing an ideal of life derived from Renaissance Italy. The ideal was called *gentility,* and it involved painstaking self-discipline and tutoring to replace the raw, "clownish" self with a refined one. Genteel living meant acquiring a graceful personal bearing and a polished style of interacting with others. By their dress, demeanor, conversation, and cultural accomplishments, genteel Americans differentiated themselves from their more common neighbors.[14]

In the course of the eighteenth century, the adoption of this ideal had generated new social occasions—tea parties, dinner parties, balls and assemblies—in which

the cultivated self might be experienced and displayed. Equally important, it had generated new physical environments to serve as settings for such occasions: spacious country houses and townhouses with ballrooms, dining rooms, and other areas primarily meant to accommodate social gatherings rather than private family interactions. Up and down the Atlantic seaboard, the mid-eighteenth century saw the construction of *gentlemen's houses* and, in many cases, surrounding gardens. Like rooms for entertainment within these Georgian mansions, the gardens outside represented ordered settings for social occasions uniting select groups of social equals or near-equals, backdrops that might accommodate and encourage genteel interaction.[15]

Like eighteenth-century gentlepeople, elite Chicagoans of the nineteenth century saw a carefully ordered environment as both the reflection and promoter of an ordered mind and character—as well as an appropriate and even necessary setting for harmonious interaction among social peers. Private horticulture in Chicago, in other words, reflected an urbane, polished ideal as much as a simple love of nature and nostalgia for rural life.

American horticultural writing of the 1840s, under the leadership of Andrew Jackson Downing, articulated aspects of this class ideal. To Downing, horticulture stood as one of the fine arts, an essential element in an American "community of rational enjoyments."[16] Starting in 1841, with his *Treatise on the Theory and Practice of Landscape Gardening, Adapted to North America, with a View to the Improvement of Country Residences,* Downing crusaded for landscape gardening and improved domestic architecture. His promotion of private horticultural pursuits appealed directly to Americans who, like Ogden, worried about the moral failings of their own and their neighbors' character in a world of nineteenth-century business pursuits. Downing insisted that the "taste for ornamental gardening" offered proof of the "progress of refinement": "It cannot be denied, that the tasteful [landscape] improvement of a country residence is both one of the most agreeable and the most natural recreations that can occupy a cultivated mind. With all the interest, and, to many, all the excitement of the more seductive amusements of society, it has the incalculable advantage of fostering only the purest feelings, and (unlike many other occupations of businessmen) refining, instead of hardening the heart."[17]

Downing influenced urban and suburban architecture in America, but he always favored country districts, which gave freer reign to his art and excluded scenes of debased urban society. He continually addressed his ideas to the country gentleman who had "wrung out of the nervous hand of commerce enough means to enable him to realize his ideal of the 'retired life' of an American landed proprietor."[18] Ogden and other Chicago gardeners approximated that life and shared Downing's ideals in only a limited way. The city's gentleman horticulturists may have valued the cultivated landscape for its freedom from commerce, but they also used it as an integral element of a larger strategy of real estate and urban promotion. After all, Ogden's appreciation of gardening and the culture such appreciation represented helped boost his fortunes. His business success in Chicago depended upon his ability to convince easterners to invest money in the city. His house, his garden, and his manner all helped him to persuade people to invest in Chicago on the faith that it would develop into a great, prosperous metropolis, one affording its elite residents the leisure to pursue culture and refinement. Ogden did not simply entertain dignitaries at his house; he entertained many investors who took measure of Chicago in part by taking measure of Ogden and his home.

In 1866, Ogden moved to New York and purchased a villa on 110 acres of Fordham Heights, overlooking the Harlem River. He gave this country seat the romantic name of "Boscobel." Jacob Weidenmann, a Swiss-born landscape gardener and a one-time associate of Frederick Law Olmsted, executed the landscape plan and natural designs for the estate. The irony of Ogden's approximation of Downing's country gentleman was that he moved to New York to pursue his business interests—among other activities, assuming the presidency of the newly formed Union Pacific Railroad. Here again, Ogden forged a link between personal refinement, evidenced in part by his landscaped estate, and business integrity. He warned the Union Pacific's board of commissioners that they must seek investments in the railroad on national and patriotic grounds, apart from all "taint of speculation."[19]

Nothing so aptly illustrates the problematic stance of William Ogden and his class as this effort to combine self-interest and its transcendence uneasily in a single moment. Chicago's gentleman horticulturists were not, after all, eighteenth-century aristocrats but speculators and entrepreneurs. Gentility and wealth fit together hand in glove, but the genteel ideal comported less well with a frantic striving after riches. Well-to-do, nineteenth-century Chicagoans modified lives of aggressive profit taking by pursuing cultivation—a means of claiming integrity, of distancing themselves not from the city but from the unbridled pursuit of Mammon. Adapting Downing's designs for country living to urban and suburban

residences and gardens gave visual expression to the elite's commitment to business integrity in an expansive, capitalist marketplace. Not surprisingly, when Ogden and his peers looked beyond their own social circles, they recommended ornamental horticulture to ennoble the lives and characters of other classes in Chicago. In the antebellum movement for public grounds, they turned from the confines of their gardens to the larger landscape of the metropolis.

Antebellum Public Grounds

Chicago's earliest public grounds drew upon much the same vision and cultural sensibility that informed private elite horticultural pursuits. Planners envisioned these public grounds—Public Square, Dearborn Park, and Lake Front Park—as an important means of displaying and diffusing social refinement and cultivation, as a means of promoting urban development and real estate values, and paradoxically, as creating a realm apart from the dominance of business engagements. The elite organization of private space found its parallel in the conception of public grounds for public leisure and recreation.

In 1837, in a move calculated to "tickle the vanity" of its four thousand residents, Chicago obtained a new charter from the state legislature that raised its status from a town to a city. Not complacent with this nominal change, editorialists urged the city to draw up a "chief list of improvements" in order to promote the "bright realities of the future." Paved streets, flagged sidewalks, new bridges, fine buildings that embodied "taste and beauty," and a municipal water system were required. The *Chicago American* insisted that the city would also need the "ornamental addition of trees which give even the most compact cities an aspect of rural simplicity."[20]

That Chicagoans desired "rural simplicity" even as they energetically promoted urban growth reveals some of the complexity in their view of their city. The city motto, *Urbs in Horto,* referred to a "garden" in the hinterlands rather than on city lots, a garden on which Chicago's existence depended. As entrepreneurs such as Ogden, Scammon, Butterfield, and Egan directed canals, railroads, agricultural implements, and immigrants into the surrounding garden, they amassed riches and built up their city. Chicago itself represented the "depot for the rich and varied productions" of the fertile West, drawing wealth from the countryside while differentiating itself from it.[21] Given this relation between the metropolis and its hinterland, what did it mean to bring ornamental trees and other symbols of rural simplicity into the depot itself? No doubt many urbanites genuinely perceived such symbols as ornamental; here then was an aesthetic in which nature seemed beautiful when possessed—not as nature, but as artifice. The *Chicago American* sought to promote not rural simplicity but "aspects" or intimations of it, intimations that truly at one and the same time seemed to testify to the absence of commerce, its presence, and the transcendence of its harmful effects on individual and social character. Ornamental trees might have represented rural simplicity; as an expenditure, they also represented the city's prosperity; as appropriated nature, they represented the source of the city's wealth in the surrounding countryside.

Many of Chicago's early public grounds took shape as integral parts of much broader plans for the development of the city; they were amenities meant primarily to boost the value of city real estate. In 1830, the canal commissioners of the Illinois and Michigan Canal initially laid out Chicago's private lots and public grounds. Charged with the construction of the canal and armed with roughly 284,000 acres granted by the United States government, the commissioners set about to capitalize on speculation (fig. 2). Selling the land would support canal construction and pay the interest on construction bonds; at the same time, the land had little value before the completion of the canal. As the transfer point between canal and lake transport, Chicago could be expected to develop as a city. More valuable than farmland, city real estate would sell by the lot and the foot·rather than by section and acre. The commission facilitated speculation by drawing up a grid plan, departing from this regular geometry only at those points where the Chicago River cut through the townsite.

The commissioners designated block 39 on the surveyor's plan as a public square. Although they did not record their vision for the use of the square, as an open public ground, or as a site for public buildings, block 39 stood apart from the speculative swirl that the canal commissioners hoped other blocks of land would attract (fig. 3). For prospective investors, the presence of a public square in the Chicago plan could conjure up metropolitan images of a city's public institutions and public life. It could recommend the city to would-be residents as well. Nothing, asserted a local newspaper, "affords better evidence of liberality and elevated taste than ample and well arranged public grounds."[22]

A single block in the center of the prospective city hardly represented an "ample" public realm, but it reflected the widely held assumption that a city would need public institutions and public life. Many Chicagoans

assumed as well that the setting for that life should contrast dramatically with the blocks devoted to private business all around it. Such a contrast would be created visually by landscaping: In the center of Chicago's booming townsite, a public square tastefully ornamented with trees and shrubs would be marked off from the business activities and structures of the city. Some residents favored restricting building on the square, barring "any structures save those that like the temple of old, rise without builders, or trowel, or hammer—the trees from the forest . . . a beautiful picture, in a rough heavy setting of brick and stone."[23] In the 1850s, when a combined city hall and courthouse building was constructed in the center of the block, it was surrounded by extensive areas of shrubs and ornamental flower gardens. According to some, such a visual contrast could affect the mood and pace of city residents. A landscaped public square might cause people to "linger" to contemplate beauty in the midst of the distracted "hurry through thronged streets" that was common elsewhere in the city.[24]

In 1835, Chicago residents sought a more substantial public park. It was unabashedly if not solely a matter of real estate promotion. The land in question was the old Fort Dearborn Military Reservation, a fifty-four-acre tract occupying the lakefront between Madison Street and the Chicago River, adjacent to the canal commissioners' plat of the town. Treaties between the United States and the Sac and Fox Indians and the permanent settlement of Chicago had reduced the importance of a military presence. The reservation now occupied valuable land in a growing city. In 1835, leading Chicago residents petitioned the federal government for the creation of a large public ground on the shores of Lake Michigan. In seeking public grounds here, they recognized, as did all subsequent park promoters, that Chicago's lakeside lands offered the city's greatest scenic amenity. The citizens' resolution asked that twenty acres be devoted to a public ground on the lake to be "accessible at all times to the people." They proposed that the remaining thirty-four acres be sold and the proceeds used for various improvement projects.[25] The *American* predicted that the park "cannot fail to gain hearty approbation, as it will be an ornament to the place, and one greatly needed, since the town is at present without anything of the kind."[26]

Many Chicago residents did give the plan their "hearty approbation." The public grounds would add to the amenities of the town; additional land central to the townsite would bring more investments to the city; and improvements funded with money from the land sale would lower taxes. Yet by 1839, when the United States Supreme Court made possible the realization of the plan

2. Land grant map and survey for Illinois and Michigan Canal,
1829 (1854 copy). The dark squares represent the sections of
land granted by the United States government to Illinois. The
sale of the land went to support the construction fund for the
canal. National Archives, Washington, D.C.

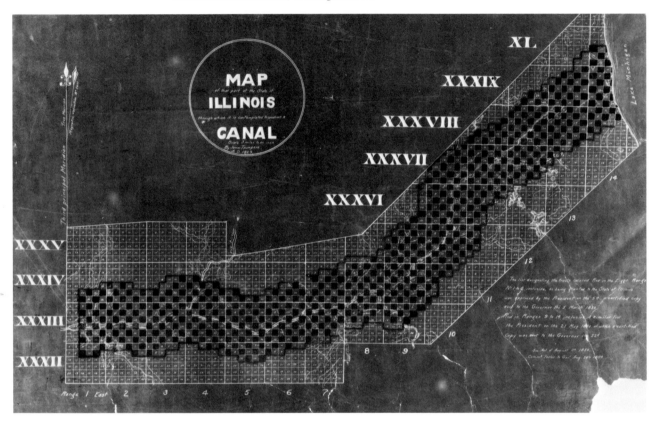

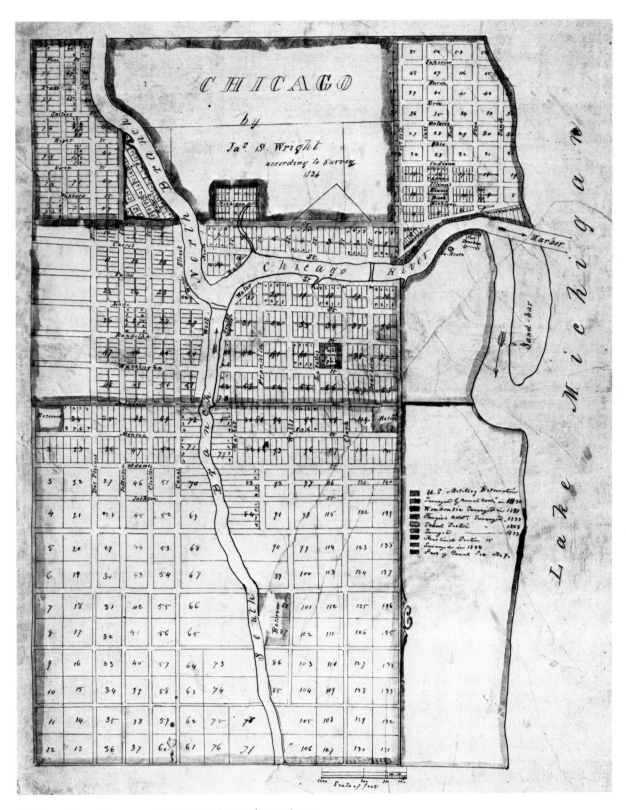

3. Chicago town site survey, James S. Wright, 1834. Map shows
the canal commissioners' original 1830 town site located on
the central fifty-eight blocks. Public Square occupies block 39;
the additions made to the town site between 1830 and 1834
are also shown. National Archives, Washington, D.C.

by setting aside a squatter's claim to the land, changed conditions in the real estate market prompted a quite different view.[27] The economic crisis of the late 1830s had burst the speculative bubble, deflating the price of Chicago land. Adding more land to the townsite would further depress an extremely depressed real estate market. Public ground or no public ground, then, many Chicago residents strongly resisted Secretary of War Joel R. Poinsett's plan to sell the Fort Dearborn Reservation.

In a petition to President Van Buren, Chicagoans declared that "unusual pressure in the money market," large land purchases recently made by Chicago residents, and the faltering of canal contracts meant that many "worthy and responsible" citizens could not participate in the government sale. "Foreign capitalists" would reap "the fruits of the Industry & enterprise of those who have rendered the reservation valuable by their adjacent improvements."[28] When the secretary of war proceeded with the sale anyway, the local Whig newspaper charged the Van Buren administration with pursuing corrupt plans to permit eastern speculators, including Charles Butler, brother of the politically powerful former attorney general, to gain control of the land cheaply.[29] The charge had some plausibility: Butler was well represented in Chicago by his brother-in-law William Butler Ogden.

In the midst of political acrimony and charges that land speculation had fostered private and public corruption, the government's creation of two public parks as part of the Fort Dearborn sale carried special importance. The government in effect met the earlier request for public grounds that were accessible at all times. The parks also preempted some of the government land from sale and market competition with Chicago's existing subdivisions. The government designated the lakefront between Madison and Randolph streets, east of Michigan Avenue, as public grounds; it also created Dearborn Park, a two-acre public square, south of Randolph and just west of Michigan Avenue (fig. 4). The land sold into private hands, for a total of $106,042, was the most valuable acreage, adjacent to the Chicago River, which served as the city's early harbor.[30] Still, a government agent confided to the secretary of war that "the vast majority of the people" appeared satisfied and the "boisterous" opponents had "seen all their lies & slander & clamor exposed & proven groundless."[31] The public grounds helped turn attention from the issues of real estate speculation. As far as the broader public was concerned, the sale of public land was apparently made more palatable by the creation of the parks. Such use of

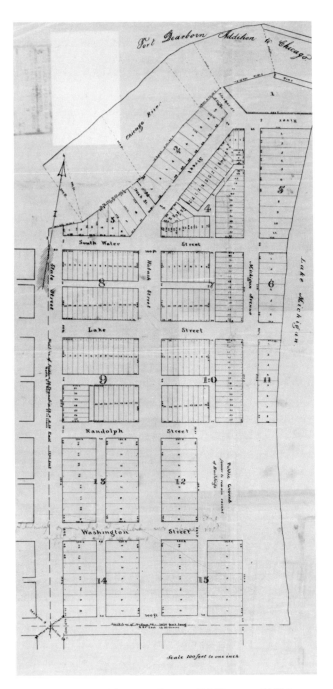

4. Fort Dearborn addition to Chicago (1888 copy of 1839 survey). National Archives, Washington, D.C.

parks to create a certain civic unity and ameliorate social and political tensions would become an enduring theme in the subsequent park movement.

The federal government's creation of a park on the Lake Michigan shore had extended a second public ground created as part of the Illinois and Michigan canal commissioners' 1836 addition to Chicago's townsite plat. On this plat the commissioners designated the entire lakefront from Madison south to Twelfth Street and east to Michigan Avenue, a "Public Ground—A Common to Remain Forever Open, Clear, and Free of Any Building, or Other Obstruction Whatever." The canal commissioners' designation clearly recognized the scenic import of the lake and thus served as both the physical and artistic cornerstone of much of Chicago's subsequent park development. In 1848, making subdivisions on Chicago's west side, the commissioners established $5\frac{1}{2}$ acre Jefferson Park.

In the 1840s, as Chicago's growth and real estate appreciation resumed, private parties increasingly dominated the subdivision and sale of city land. Some private developers followed the precedent set by the canal commissioners, dedicating small squares to the city for public parks in the midst of blocks and lots offered for private sale. Real estate dealers anticipated that public embellishments of these parks would enhance the value and beauty of adjacent real estate. Bushnell's 1842 addition to Chicago, for example, deeded the $2\frac{1}{4}$ acre Washington Square to the city with the condition that it be fenced and improved. In the 1850s, Henry D. Gilpin deeded the four-acre Vernon Square to the city in connection with his own west-side real estate development.

In the 1850s on Chicago's north side, Samuel H. Kerfoot developed a private park which he opened to the public. Kerfoot was one of Chicago's most successful real estate investors and developers. His park stood in the midst of land that he was promoting as an elite residential section five miles north of Chicago's downtown. Kerfoot, in his own account, sought to create the "first specimen of artistic landscape gardening in this section of the country . . . [the] most thorough piece of work in its way west of the Hudson River." The ten-acre park, later developed as Kerfoot's residence, included an extensive carriage drive, shaded by planted evergreens, artificial ponds crossed by bridges, and rustic arbors, steps, and seats (fig. 5). As a "gentleman" on the model of Ogden and other early private horticulturists, Kerfoot kept up his garden and maintained one of the city's finest private libraries and collections of paintings. A biographical notice on Kerfoot mentioned his landscape design and

5. Kerfoot's Park (photo ca. 1864–69). In the 1850s Samuel H. Kerfoot opened this ten-acre park to the public on land he owned on the north side of Chicago. The park helped promote Kerfoot's surrounding land for expensive residence lots. Chicago Historical Society, ichi 22331.

suggested something of the impetus behind the local park movement: "With real estate is naturally connected a love of horticulture, arboriculture, and landscape gardening. In this particular Mr. Kerfoot has indulged extensively, and has shown great taste and skill."[32] In this formulation from the 1860s, speculation and horticulture did not stand opposed as social ill and antidote; Kerfoot's biographer noted a similarity in their stance toward nature unimproved, a shared interest in appropriation of the land, expansion, and what could be seen as the creation of something from nothing.

In the transition from private gardens to public grounds, the social conception of ornamental horticulture changed. Proponents of public horticulture suggested that society, not merely individual gentlemen, would benefit from cultivation. Of course, gentlemen horticulturists could view their own gardens as a means for improving society. To the extent that they tempered avarice in themselves, they served the social whole. Moreover, the *Journal* suggested in 1847 that many residents shared the enjoyment of gardens that adorned estates of the city elite: "The gush of fragrance from the flowers, borne on some passing breeze, comes like a sweet whisper of hope to the poorest pedestrian as he returns from toil, and the beauty of the shrubbery he shares in common with all." Yet regardless of the extent of public benefit gained from these "ornaments to the city," private gardens remained private, "enduring monuments" to the "good taste" of their owners.[33] Although some modest residences could boast flowers, trees, and shrubs, most Chicagoans lacked gardens of their own. Public grounds served a broader constituency and were envisioned by their promoters as effecting a broader basis for social refinement. Public grounds would be accessible to all.

Unlike gentlemen gardeners on their own estates, most members of the public would not actively cultivate trees, shrubs, and flowers on the city's public grounds. Yet park promoters hoped that ordinary Chicagoans might still cultivate their own characters—first through uplifting contact with natural scenes and, second, through the more gregarious activity of promenading. Many elite Americans were first impressed with the promenade custom as they encountered it in Europe. William Cullen Bryant, an antebellum leader in the movement for public parks, for example, had observed it during a visit to Munich. He reported that after Sunday religious services residents "proceed to some of the public walks and gardens and amuse themselves by walking about, observing the crowd and greeting their acquaintances."[34] Antebellum Chicagoans followed a similar practice. In the 1840s, the Lake Front Park emerged as the most popular area of public promenading in the city. After the fashion of early-nineteenth-century promenades in Europe, along New York's Battery Park, Boston's Tremont Street Mall, Philadelphia's Fairmount Park, Charleston's Battery, and on the edge of Brooklyn Heights, Michigan Avenue and the adjacent public grounds (fig. 6) developed as "the Battery of Chicago."[35] Amid the locust trees on the lake side of the avenue one could find on Sunday evenings "crowds of well-dressed people . . . promenading up and down." Families with "toddling" children mixed with young couples, "heads and hearts full of the one great theme of earth and heaven."[36] Promenading was an occasion for people to see and be seen, for the decorous display of fashionable dress, graceful demeanor, and social polish.[37]

Elite enthusiasm for the promenade custom lay primarily in its supposed ability to unite different classes of the city in a common activity, creating a public representation of social harmony. In a city noted for the feverish pursuit of profit, the possibility of strife between classes loomed. With scenes of commerce and debased character largely absent and pictures of pure, moral nature present, the promenade offered the benign vision of a coherent social order coalescing around shared leisure and shared urban space. In 1830 Benjamin Silliman's *American Journal of Science* asserted that with more public grounds and promenades in American cities, the "feelings of th[e] people will flow in a kinder and smoother channel." Such an institution could counter the "fondness for change," the "restlessness," the "greater individuality," and the "pressure of business" that some writers thought pervaded American character.[38] Promenading, claimed another proponent, would "tend to bring the classes of society into a more intimate relation, thus happily doing away with many of those petty jealousies and distinctions that unhappily exist."[39] Similarly, Downing argued that the artificial "barriers" of class, wealth, and fashion erected by Americans could be modified by the creation of public parks, gardens, and galleries, a "community of rational enjoyments."[40] The tensions produced in the realm of work, in other words, might be allayed in a realm of leisure. Classes might mix on the promenade, "a ground resorted to by [the] whole population for air, exercise, and recreation . . . where the poorest man in the city can go as unquestioned an owner as the richest."[41]

It seems unlikely that the poorest Chicagoans did attend the promenade. In a world of fine social distinctions, commentators' testimony that all classes participated might best be taken to indicate the presence of a

6. Lakefront Park and Michigan Avenue looking northward
from Congress Street (photo ca. 1868–69). The trees planted
along the fenced sidewalk of Lakefront Park, Chicago's primary
promenade area before the creation of the 1869 park system,
are visible in this view. Starting in the 1850s, the Illinois Central ran along the track on the trestle in the water. Grain
elevators are at the right in the distance. Chicago Historical
Society stereograph, ICHI 04438.

range of urban dwellers mostly from the middle and upper classes.[42] For all that, the fact of class mixing was notable. Promenading was not merely an elite custom, and it followed that support for public grounds designed for the promenade was not limited to the upper classes of the city. Other members of the emerging middle class, as well as the most wealthy Chicagoans, backed the creation of public grounds.

The middle class's enthusiasm for public parks and the promenade custom represented in part its members' appropriation of the ideal of cultivation. In the early nineteenth century, middle-class Americans took into their own lives the quest for personal refinement and social grace. Culturally, it was this commitment to the cultivated self that gave the middle class its distinctive identity. "Respectability"—a less aristocratic ideal than gentility—became its hallmark as a class. The public promenade allowed a presentation of the respectable self; and although one commentator characterized the promenade custom as partaking of "nobility," such self-presentation was increasingly central to the aspirations of America's middle class (see plates 2 and 3).[43]

As much as the supposedly ennobling effects of nature, then, the institution of promenading made sense of the idea that parks might unify a fractious city society. Indeed, the movement for public grounds and promenades did function to unite Chicagoans of different economic strata. An elite and middle-class alliance emerged in the movement for parks. It was an alliance forged in large part around a shared understanding of the need for and nature of urban cultivation. Chicagoans elaborated their concept of urban life and landscape as they joined together in—or took up opposition to—the postbellum movement for an extensive park system.

Chicago's Park System

In the late 1840s and 1850s, Chicago officials began to approach the problem of public grounds more systematically. In 1847, Mayor James Curtiss proposed that new parks of from ten to twenty acres be established in each of the city's three geographical divisions. He declared that parks would "furnish an inexhaustible fund of pure and refined gratification."[44] Pushed by considerations of urban pride, public expectations, an expanding area and population, burgeoning horticultural interests, and visions of a "densely populated" city, the Common Council purchased land in 1854 for the first of the division's parks. The council established Union Park, an eleven-acre ground on the city's west side. It was the first park obtained by the city through its bonding and purchase

authority rather than through free dedication, and the city's appropriation for it affirmed municipal responsibility for the provision of park lands.[45] Establishing a precedent followed in more ambitious park purchases, the city specially assessed property nearby to help defray the costs and avoid charges of inequity in public improvements (fig. 7). The special assessment system codified the understanding among real estate dealers that parks raised the value of adjacent property specifically and of city property generally.

Beyond tentative efforts to lay out parks in the city's three divisions, the late 1840s brought the most perceptive forecast of Chicago's future park system. In 1849, John S. Wright, one of Chicago's most energetic boosters and real estate developers, wrote "I foresee a time, not very distant when Chicago will need for its fast increasing population a park or parks in each division. Of these parks I have a vision. They are all improved and connected with a wide avenue, extending to and along the lake shore."[46] Such a grand vision for Chicago parks lay fallow through the economic crisis of the late 1850s and the political crisis of the early 1860s. In the late 1860s, the horticultural achievements of Chicago's private gardeners, the early initiatives in public grounds and promenades, and a continuing belief in the moral and economic importance of landscape design coalesced in a permanent park system on the order of Wright's vision.

"A GIGANTIC IMPROVEMENT" declared the *Chicago Times* in September 1866, when it published a plan for Chicago parks and boulevards that was fully equal to

7. Union Park, ca. 1864–70 (photo ca. 1864–69). Chicago Historical Society, ICHI 22333.

Wright's earlier vision. In a fairly unusual venture into graphic reproduction, the *Times* showed a cartographic cut of the proposal: a park one-quarter of a mile wide and fourteen miles long, completely encircling Chicago from the southern lake shore to the northern lake shore, at about the city limits. Boulevards lined with land divided into building lots flanked either side of the continuous park strip (fig. 8). The plan nicely fitted the pattern of confident, optimistic, urban boosterism pervading Chicago's business and civic culture. The *Times* argued that the improvement would beautify Chicago, giving it "the finest drives, parks, and building sites on the continent."[47] The 1866 park plan was city building on a grand scale—so grand that the *Times* suggested that its "vastness" might militate against its prompt realization. Nevertheless, within three years Chicago had, in a somewhat modified form, a park system of similar extent and intent. Between plan and reality the number of acres declined from 2,240 to about 1,800; the continuous park strip was grouped into distinct park areas connected one to the other by landscaped boulevards, and the public building-lot development anticipated in the earlier plan became a private and implicit part of the plans adopted in 1869. These changes aside, within a very few years Chicagoans realized a substantial reconfiguration of their city.

Historians have generally explained the creation of such extensive park systems as a manifestation of an American *antiurbanism,* a nostalgic response to passing rural life. Galen Cranz, for example, asserts that the park movement "derived not from European models but from an antiurban ideal that dwelt on the traditional prescription for relief from the evils of the city—to escape to the country." Lewis Mumford characterizes parks similarly, as "a means of escape,"—a "refuge against the soiled and bedraggled works of man's creation." Ross L. Miller concludes that landscape architects, embracing "Jeffersonian notions," and wedded to "cultural ideas associated with a pre-urban society," sought in park designs to impose "older pastoral values, three-dimensionally, in the form of park lands, upon a basically hostile cityscape." More recently, Thomas Bender has argued that nineteenth-century Americans simultaneously sought a closeness with nature and civilization. Parks, he suggests, represented the convergence of these two impulses in a form of "middle landscape." Yet, like other park movement historians, Bender portrays nineteenth-century experience of the city as largely negative. Urbanites, he implies, found city life temporarily tolerable largely because of the tempering influence of nature and its associations with the Garden of Eden, God's immanence, and rural virtue and purity.[48]

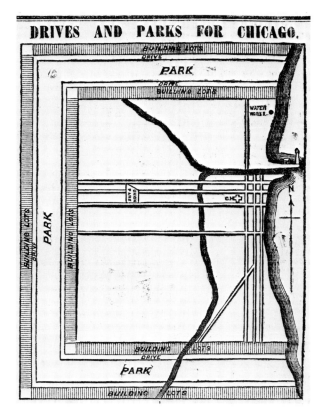

8. Unexecuted plan for Chicago drives and parks, 1866. From the *Chicago Times,* 15 September 1866. Chicago Historical Society.

Indeed, park history contains an antiurban dimension expressed in the literature of park promotion and other nineteenth-century forms. Like many twentieth-century historians, some nineteenth-century Americans assumed an irreconcilable opposition of city and country, art and wilderness, and civilization and nature. Drawing on both European romanticism and American nationalism, many prominent nineteenth-century artists and intellectuals assigned greater moral, aesthetic, and political value to various forms of nature than to human society and its creations.[49] Yet this is only one aspect of park history (fig. 9); taken by itself it misrepresents the interests and impulses behind the park movement, which constituted a more serious attempt to shape urban realities than the nostalgia theory admits.

As Chicagoans pressed for a larger park system, they elaborated upon assumptions that had informed the cultivation of private gardens and the early movement for public grounds. Parks, many believed, might create a literal and symbolic distance from commerce; help establish the cultural ascendancy of bourgeois values; create stable single-class residential areas; and, in apparent contradiction, formulate a cross-class community of leisure. In the park movement, some Chicagoans used the landscape to express and form social identity and social relations (fig. 10). For many these goals were inseparable from that of constructing their city.

In 1874, Horace William Shaler Cleveland noted that people's interest in parks followed a deliberate removal from agrarian life, a removal spurred by a desire for all the "artificial elements of refined civilization which can only be had in cities." It was "a very curious fact," he noted, that a desire for "refinement of uncultivated taste" led people first to move to towns, next to build grand city halls, stores, hotels, and residences "to make a display of wealth and magnificence of architecture and artificial ornamentation," and then finally "to surround themselves with the objects of natural beauty which at first . . . seemed worthless." If this was a manifestation of nostalgia for country life (a sentiment easily indulged at a distance), it was also prompted by visions of urban preeminence. By revivifying, at great expense, objects of natural beauty that "man first destroys to make room for his own creations," city residents linked cultivation of nature with artificial patterns of social fashion, refinement, aesthetic uplift, and city building. Those elements of *agrarianism* or *nature* enjoyed in parks were several steps removed from farm and forest; in many regards these elements were as "artificial" as the rest of urban life. Moreover, park promoters easily acknowledged the judgment that such artificiality was a good thing. In

Cleveland's own mind, for example, parks and gardens represented "the best evidence" and "the most luxurious form" of "refinement and cultivated taste."[50] Urban enthusiasts of landscape gardening identified it as "one of the Fine Arts . . . capable of exhibiting a richness of beautiful effects scarcely less brilliant and various and impressive than architecture itself."[51] Like the city's examples of fine architecture and its early cultural institutions, the parks presented one more of the "luxuries and conveniences which only a great city can afford."[52] The wealthy boosters of Chicago's park system aggressively sought to meet national and international standards of cosmopolitanism, to promote real estate values, and to realize the progressive possibilities of urban culture.

Thus, Chicago residents such as Ogden, Scammon, Kinzie, Butterfield, Illinois Lieutenant Governor William Bross, and Ezra B. McCagg were also aggressive boosters of the city itself—its commerce, industry, real estate values, and ascendancy in the ranks of American cities. In many cases their standards for urbanity came from a familiarity with European cities. Ogden, traveling in Europe in 1853 and 1854, and Bross, who toured Europe in 1867 and 1868, both sent letters to Chicago papers urging the development of a number of important urban institutions, including public libraries, museums, parks, and boulevards.[53] In 1859 a Chicagoan visiting Europe, perhaps Scammon, reported being struck by the extensive parks and gardens in all European cities and concluded, "Chicago has nothing deserving the name of a park . . . [and] ought not to style herself the 'garden city,' without a garden or a park." That the city lacked the great libraries, art galleries, and cathedrals of Europe could be excused on the basis of the city's youth; however, the traveler concluded, the failure to provide for Chicago parks, "the first and most essential want of a city, when it could most easily have been provided for, is the worst charge that has ever been brought with justice against her."[54] Park development in Chicago could refute charges of civic irresponsibility and provide a precedent for other Chicago cultural institutions. As one supporter of the 1869 park plans hoped: "Parks first, and the Museums and Libraries will follow in due time."[55]

Chicago park proponents also expressed the sense that they were competing with American cities for preeminence, particularly with New York, whose Central Park made other cities "conscious of their deficiencies" and prompted them to "adorn themselves in like manner, or try to eclipse [New York] by a more costly decoration."[56] Nor were Chicagoans alone in taking eastern cities as a model. Cleveland reported that with railroads and growing emigration to the West, the residents of

9. Sheep grazing in Washington Park, ca. 1895. A useful means of keeping the grass short, the sheep in Washington Park fostered a somewhat false image of park pastoralism. Chicago Historical Society, ICHi 17003.

10. Band concert in Lincoln Park, 1907. Library of Congress.

many new towns insisted from the earliest days of settlement upon the "comforts and luxuries of civilization." A lag no longer existed in the West's adoption of the habits, dress, idioms, conveniences, and town forms of the settled East.[57] At the same time, Chicago struggled against a particular stigma, a reputation for being only "half-civilized" that fueled the determination of civic leaders to draw up plans for parks. Thus, in 1871, Olmsted expressed some surprise at the cultured existence of the family of Ezra McCagg, a wealthy Chicago lawyer and philanthropist:

> Having all migrated to the West while young & lived for a while on the very frontier & in the midst of the maddest whirl of speculation ever known, you might expect [the McCaggs] to be very different from what they are. . . . They are passionately fond of flowers & the house is always rich in them. They are all but Mr. McCagg passionately fond of music. . . . They are rather sensitive about the West & Chicago lest any one should think that people are not likely to be as well informed & cultivated there as any where.[58]

The prevailing belief that parks served as "an unerring index of the advance of a people" encouraged McCagg to become a park proponent and eventually one of the city's first park commissioners.[59] Like Chicago's private gardens, parks appealed to Chicago's gentlemen, among others, in part because they represented time and resources preempted from the individual pursuit of riches. In providing for the needs of both present and future generations, the park movement suggested the existence of an enlightened civic culture. When Olmsted visited Chicago in 1863 he found little evidence of thought for future generations; the lack of a large public park represented to him a decided poverty of public life and amenity. He was told that selfish jealousies over who would benefit from the park and the lack of "a man of ability" to lead the movement had frustrated park plans. During this visit, Olmsted urged upon his Chicago host (probably Ezra McCagg) the need to take responsibility for park development.[60] He pointed out that McCagg, in his private garden, enjoyed the benefits of trees and grass and that in his European travels had witnessed the public benefits to body, mind, and soul of large parks. So why shouldn't he lead a park movement? McCagg confessed that the pressure of legal trusts placed on him simply did not permit him the time; however, McCagg did admire the supposed European gentleman's standard of public leadership, a standard that he aspired to later when he

did eventually assume a leading role in park affairs and cultural philanthropy. He explained to Olmsted:

> Five years ago, when I returned from Europe with my family, I brought with me books, pictures, statues, which I had carefully selected for my own enjoyment, as you see I brought them to my house here, with a resolute determination that I would enjoy them, that I would live as gentlemen do in Europe. I saw what a useful and even essential part, those whom we call here idle men play in the sum of business of every well-established community, doing, in fact, just such work as . . . is unfortunately for us and our children, left undone here for want of them.[61]

This defense of aristocracy suggests the appeal of the park movement for some of the city's elite. Cultivating the city could counteract any charges of idleness leveled against those of leisured wealth; further, it justified the displacement of one's energies from the commercial world into cultural pursuits, a transition that—in the eyes of an increasingly dominant middle-class culture—appeared both difficult and particularly important.

Park proponents generally framed matters more simply, by portraying the cultured life afforded by commercial wealth as commerce's only real justification. Bross, one of McCagg's major allies in the park movement, considered it "the power of unsanctified wealth to corrupt and debase the national mind." The *sanctification* of wealth meant breaking a link between riches and "social and individual demoralization"—something accomplished only by the proper educational and benevolent use of money. Once the vision of "gigantic improvements" in the form of parks arose, the choice for Chicago's elite appeared simple:

> If Chicago is to be anything better than a hive or a rookery, where men buy and sell and get and gain; if it is to become the residence place of men of wealth, elegance and culture, and not merely a dirty and smoky resort, where merchants and manufacturers, bankers and grain dealers, and beef and pork butchers and packers make their money and then leave, we must have parks.[62]

By countering (or perhaps complementing) commerce, parks would help make the city a place where the right sort of people chose to stay (fig. 11).

The struggle surrounding Chicago's first major park legislation, passed by the Illinois legislature in 1867, revealed the extent to which these ideals coexisted with other interests in urban public grounds. Cleveland, who

11. Approach to Lincoln Park and Saint-Gaudens's Statue of Lincoln (1900–10 photo). Library of Congress.

both loved nature and believed in its refining and uplifting power, recognized the divergent sources of the park movement. Although "a leaven of genuine love of nature" existed in America, he wrote, it represented only "a very small proportion of the motive power of the world and it is very hard *anywhere* to get up force enough to give steerage way to the community. It is absolutely necessary to enlist other motives in order to accomplish anything and there in lies the despair of all who love nature for herself alone."[63]

The issue was not merely that many preferred a more gregarious use of the public space—a preference that led many park designers to incorporate promenade grounds into their designs. Most distressing to those who viewed parks as capable of ameliorating a debasing preoccupation with profit was the degree to which park promotion flamed the material passions of real estate and building interests. A public improvement that involved the expenditure of millions of dollars and the acquisition and development of hundreds of acres of urban land could hardly evolve free from pecuniary calculations. This proved especially true in a city in which many of the largest fortunes derived from real estate. In February 1869, an associate of reaper manufacturer and real estate investor Cyrus Hall McCormick wrote to McCormick about progress in the park movement. He did not mention nature love, public health, cultivation, or refinement; he wrote that the movement for parks and boulevards would "have the tendency to advance outside property." He promised to send maps of the parks so that McCormick could plan his real estate investments.[64] Predicting that the park improvements would attract at least $1,000,000 in "foreign capital" investments, benefiting both rich and poor, the *Chicago Tribune* concluded an 1869 editorial supporting the park bill by declaring: "The easiest way for every man in Chicago to make money is to vote for the parks."[65]

If the *Tribune* exaggerated, it was nonetheless clear that some Chicagoans would profit from the parks. Economic interests were so clearly at stake that the Illinois legislators who passed the 1867 park bill felt pressed to deny the influence of self-interest. The bill, outlining a park of up to a thousand acres for the south division, originated with a small group of "leading citizens" most ready to embrace the ideals of nature, refinement, and cosmopolitan attainment. Scammon, who helped write the legislation, had grown up in Maine and felt that had it not been for a disabling childhood injury to his hand he would have spent his life as a farmer.[66] A biographical note portrayed Scammon as "a scholar, of refined culture and great attainments . . . and a courteous gentleman."[67] Rev. Robert W. Patterson of the Second Presbyterian Church and another "gentleman" had drawn Scammon into the park movement by urging upon him the "necessity of immediate measures being taken to have a large Public Park." Pressed by legal business, Scammon asked his partner Ezra B. McCagg to draw up a bill based on New York's Central Park statute. Claiming the public spirit of a gentleman and statesman, Scammon denied imputations of self-interest by announcing, "there ha[s] been neither clique nor combination upon the subject."[68]

William Bross echoed Scammon's assertions of public spirit, declaring that he was "laboring wholly in the interest of posterity" in supporting the park bill. On the floor of the Illinois Senate, Bross denied charges that he held a pecuniary interest in the project. In fact Bross credited an 1865 meeting with Olmsted in Yosemite Valley with sparking his interest in the park question:

> I met Fred Law Olmsted on top of the Sierra Nevada. From those dreadful peaks . . . until we feasted with the pigtail mandarins in San Francisco, we discussed nothing so much (pardon me, with the exception of the Pacific railroad) as the Central Park of New York. And both Colfax and Olmsted agreed with me that nothing was needed to make Chicago the principal city of the Union but a great public improvement of a similarly gigantic character. My return being hastened by my ardent desire to have the thing attended to, I at once consulted the principal citizens of Chicago.[69]

After receiving a copy of Olmsted's 1866 report on a park for San Francisco, Bross conferred with Olmsted about Chicago's nascent park movement.[70]

However pure the motives for wanting a public park system, in Springfield the real estate interests turned out in force to do battle over the park's size and location. Developers naturally favored different sites for the improvement. The *Chicago Times* reported after a month of the legislative session: "there is a *Tribune* (which maybe called the Deacon Bross) park, Scammon's park, Paul Cornell's park, Ezekiel Smith's park, Horse Eddy's park, Perkin's park, Drexel's park, Bond's park, and Kerfoot's park."[71] Disagreement peaked over the designation of Dr. William B. Egan's old nursery and gardens as the center of the proposed park. From a horticultural and scenic point of view, the plan made good sense. Yet the Drexel brothers of Philadelphia and the Smith brothers of Chicago, who jointly owned the property, resisted the designation. Their land would be worth more, they believed, if developed as elite residential lots adjacent to a park rather than taken over for the park itself. The Drexels found a capable representative in lawyer and lobbyist Norman Williams. A *Chicago Tribune* correspondent reported, "The only real opposition comes from the Drexel champions, and I do not think their pile of greenbacks is large enough to defeat it."[72]

After considerable debate, the 1867 legislature ducked the issue, designating a section of the city for the park but leaving the precise location up to a commission established by the bill. Paul Cornell, a leading real estate developer of Hyde Park who maintained an active material interest in the entire south-side park question, helped engineer the park commission's decision to leave the Drexel property outside the proposed park.[73] Under the circumstances, it took considerable imagination for some of Chicago's "best citizens" to proclaim that the parks plan had "not a penny of speculation in it."[74]

The charges that swirled around the 1867 park bill were blamed for voter rejection of it in April of that year. The *Times* suggested that defeat was "mainly attributable" to the "very general opinion that no act of the recent general assembly which was capable of containing a 'steal' did not contain a 'steal.' " Voting patterns in the city suggest other considerations that affected the popular decision. Two wards of the city, the first and the fourth, voted favorably on the bill.[75] The fourth ward, the seat of the new park, would benefit most from its construction; in the downtown first ward, the *Times* opined, working-class voters who did not pay property tax chose to vote the interests of their employers who lived and owned land in the fourth ward. Elsewhere in the city, the paper believed, the rejection amounted to revolt against rising taxes by property owners, who apparently preferred lower costs to the amenities of a new park.[76]

Other elements contributed to the park's defeat. Park advocates took some pains to woo working-class support, reminding the poor that the millions of dollars spent on park improvements meant employment for thousands of laborers.[77] Before the vote, one writer had insisted that the

> masses will not be taxed for the park but only the rich, who have been made rich by the labor of the masses. It is the laboring man's measure, and they, above all others are the most interested in getting the park. They must have some place easily accessible in which to spend their leisure hours during the hot weather with their wives and children; but the rich can visit

foreign countries and places of popular resort in other states.[78]

Despite such assertions, in the 1867 election the working-class districts and wards, those that clustered along the South Branch of the Chicago River, in industrial sections, and around rail yards, voted by large majorities against the park bill while the middle- and upper-class districts and wards, those areas along Lake Michigan, voted in favor. In the second ward, for example, the middle-class residential area bounded by Monroe, Clark, Harrison streets and Lake Michigan, voted 427 for the park and 256 against. The other half of the ward, more heavily working-class, voted 94 for and 391 against.[79]

Working-class opposition to the park bill stemmed from a variety of sources. Working Chicagoans were not immune to property taxes. Some of them owned property or hoped to do so; others no doubt considered the probable effects that new taxes would have on rents. It was true that taxes would fall most heavily on property in other parts of the city, since special assessments concentrated on the parks' adjacent residential land. Under this system, one park proponent noted, " 'Those who dance must pay the fiddler.' "[80] Yet that merely underscored the fact that the parks would not be located in working-class neighborhoods, whose residents would "dance" little if at all.

In 1869 the legislature passed bills better adapted to securing public ratification. They established major landscaped parks at lakefront sites in the north and south divisions, four parks inland from the lake, and a system of landscaped boulevards connecting all of the parks into a circuit built at a distance of $2\frac{1}{2}$ to 6 miles from the Court House on public square. This distribution of park space was intended to broaden public approval, and Chicago voters did ratify the bills in 1869. As had happened two years earlier, the strongest support for the park movement came from Chicago's middle and upper classes. Working-class districts again voted against the park bills, although by smaller margins than in 1867.[81] The new plan still left many laboring Chicagoans far from a public park.

Indeed, a central aspect of the park system was its removal from the older, settled area of the downtown. In public discussion of the issue, park advocates had sung the advantages of parks for residents interested in such extra-park matters as transit lines, residential development, and a more orderly pattern of industry and commerce. Parks, they promised, would provide the framework for the development of residential neighborhoods protected from the "invasion" or "intrusion" of

12. Lake Shore Drive, looking north toward Lincoln Park (1905 photo). Lake Shore Drive stood out among Chicago's residential boulevards by combining the usual boulevard amenities with a direct view onto Lake Michigan. Potter Palmer's 1882 mansion stands at the left. Library of Congress.

industrial, commercial, or tenement buildings. The connection envisioned between parks and the residences of the "wealthier" classes emerged as something of an unquestioned assumption of park proposals (fig. 12). The social ideal of cultivation and refinement expressed in the earlier development of gardens and public promenades was largely a reflection of these classes' sensibility. The upper class in particular enjoyed ample leisure for using the parks and at the same time most readily affirmed their importance in the urban social world. Real estate interests assumed, in many cases correctly, that upper-class interest in parks would result in the development of wealthier residential sections in close proximity to the newly formed parks and boulevards.

Chicago's 1869 park bills by and large established parks beyond the settled areas of the city, in places where commercial and industrial encroachment did not present an immediate threat. In the midst of Chicago's booming post–Civil War growth, then, the park plans represented an act of speculative optimism which assumed the city would quickly overtake and surround them. Park areas, excluding commercial associations and complementing domestic sensibilities, seemed to offer the most promising foothold for residential development. Important European park precedents suggested the possibility of linking park landscapes with upper-class neighborhoods set apart from commercial and industrial sites. Such major European parks as Birkenhead, St. James, Regents, and Bois de Boulogne, which influenced American park designs, included residential sections as an integral part of their planning and financing. In Europe the profits realized from the sale of houses at the park periphery helped fund park improvements.[82] Under sway of a more laissez-faire, decentralized free-hold system of urban land development, Americans modified this practice, leaving land adjacent to parks and boulevards in private hands but taxing it highly. If differential taxation made matters fairer, it also meant that only families of some substance could afford to live in the area.[83]

Chicago's flat, uniform topography exacerbated the difficulties of effecting residential stability. Apart from the lake shore, Chicago's builders and residents found few obvious natural settings for "elegant and costly private houses."[84] In the midst of such relative uncertainty parks served as magnets for residential development. Transportation lines were readily extended to them. In late-nineteenth-century city atlases the patches of green parks and boulevards loomed large in relation to the monochromatic portrayals of the surrounding areas (see

plate 1). One writer declared that despite the fact that Chicago occupied a "vast plain" residents would not settle it "indiscriminately. . . . If the 'plain' is anywhere relieved by flood, forest, or hill, population will be attracted thither."[85] In their formal design features parks and boulevards represented a powerful clustering of flood, forest, and hill. The parks and boulevards clearly spurred greater real estate appreciation and more substantial residential character than most nonpark areas.[86] In sum, parks and boulevards created a world away from the downtown, providing for some well-to-do residents an alternative center for urban life.

What spurred much middle-class and elite support for the 1869 park bills was competition among different divisions of the city for wealthy residential areas. The legislation appeared to project a unified park system, based upon a comprehensive vision. The continuous park and boulevard system created a landscaped frame or boundary to the city which played an important role in the practical and imaginary ordering of Chicago's landscape. Yet despite the suggestion of a unified system, the park plan ironically emerged from the strenuous competition between Chicago's three geographically defined divisions. The park bills created the South Park, West Park, and Lincoln Park commissions, completely separate and autonomous bodies. The parochial competition between divisions and property owners' fears of being eclipsed by other divisions added momentum to the park movement.

In 1871 the West Park Commission, for example, maintained that park improvement would "attract hither thousands of people of leisure and culture, whose presence among us must add much to the material as well as social wealth of the city."[87] Park supporters charged that a small clique of real estate owners from the north side, which did not need voter ratification of its park bill, was behind the opposition to the bills establishing parks in the west and the south. This opposition amounted to "a very heavy blow aimed directly at the superior advantages and prosperity" of other parts of the city.[88] On the north side real estate values were "beneficially or disastrously affected just as Lincoln Park [was] advanced or retarded"; as one adviser explained to a large property owner, "The active rivalry between the various Divisions of the City in securing parks & public grounds & boulevards, and thereby attracting the wealthier & more desirable classes of population . . . seems to decide very clearly what the policy of the North Side is."[89] Failure to proceed with park development would render residential property only second- and third-rate, less profitable for investors and incapable of attracting wealthy residents. Thus, in the 1860s the real estate interests, the civic leadership, and the voters of Chicago's three divisions competed for the amenity and status of large public parks as a way of competing for elite residential neighborhoods.

Residences away from the downtown appealed to middle-class and upper-class families who increasingly envisioned a city that was segmented spatially, socially, and culturally. Enhanced by improvements in urban transportation, building technology, and changes in the structure of business life and hours, residence increasingly separated from business quarters in growing nineteenth-century American cities. Ascendant middle-class domestic and cultural ideals favored settings somewhat removed from the associations of Mammon.[90] Chicago's nineteenth-century growth often suggested to observers the "magical creations" of Aladdin's Lamp.[91] Although they often took pride in such growth, upper- and middle-class residents alike found aspects of the City's expansion disconcerting. As one writer noted, for residents who believed that their neighborhood would "be forever dedicated to the uses of domestic and home life, this encroaching advance of commerce, with its obliteration of old scenes, is regarded with anything but feelings of pleasure."[92] Residents unsettled by commercial and industrial expansion cast concerned eyes across Chicago's undeveloped landscape in search of more stable settings for their residences. In both understanding and directing broad patterns of urban development, then, park planning went far beyond the functional and recreational role assigned to it by most historians of urban planning.[93] Park planning was, quite self-consciously, tantamount to city planning.[94]

Indeed, beyond providing a framework for wealthy residential sections, the park and boulevard system helped define a new physical and metaphorical boundary for the city. Again the topography suggested only limitless, undirected growth. In 1863, Olmsted recorded in his journal that, "It is sad to see with how little forethought the town is nevertheless suffered to enlarge. It is only a multiplication of parallelograms upon a flat surface."[95] Surrounding the city, parks and boulevards gestured toward future growth. However, they also set up a fundamental distinction between "inside" and "outside" land and suggested a reassuring sense of the city's limits. The cordon of parks and boulevards (see plate 1) would surround Chicago "as the old wall surrounded the Roman cities . . . attracting the attention of persons looking

for eligible residence sites."[96] In 1866 the Chicago Common Council objected to the fourteen-mile circuit park plan because it appeared to confine the city within too narrow limits.[97] The 1869 park bills avoided the criticism of narrow bounds simply by outlining a broader container.

The perceived scale of a city bounded by a boulevard and park system contrasted sharply with a city laid out in a street grid. Many felt that Chicago's grid design, with few real distinctions between blocks, was implicated in the rapid encroachment of commerce and industry on domestic quarters. After residing in the Midwest for a short time, H. W. S. Cleveland grew convinced of the "urgent importance of making a vigorous protest" against the prevailing "rectangular ideas" of town planning.[98] In 1873, he recommended a new form for the city neighborhoods which drew upon the concept of a landscape frame established by the broader park system:

> Let us suppose the central and most important business portion of the city to be surrounded by a series of small parks, connected by broad avenues or boulevards, tastefully planted and adorned with fountains, flower beds and appropriate works of art. Let other portions of the city, appropriated to special branches of business or manufactures, be similarly surrounded and isolated, and from each of these areas let a series of boulevards radiate on lines diagonal to the general course of the streets.[99]

The landscape architect's framework would supersede the engineer's grid (fig. 13).

Cleveland's plan for commerce and manufacture bounded, surrounded, and isolated by radiating avenues and landscaped settings had obvious appeal for wealthy urban residents. It gave landscape form to the familiar apology for the commercial city which accepted a commercial kernel in connection with a cultured stock; here commerce had a prescribed position that did not threaten the proprieties of Chicago's domestic, leisure, religious, or cultural pursuits. Cleveland viewed his boulevards as facilitating the full enjoyment of theaters, museums, libraries, lectures and social gatherings by permitting access between these institutions and residences not "marred by an association of physical discomfort," including the discordant contrasts of commercial and industrial life.[100] Although traditional land divisions and prevailing business practices militated against Cleveland's plan, the proposal concisely articulated central features of the park system's broader relation to urban patterns and attitudes. Both Cleveland's design ideas and

13. Looking south along Drexel Boulevard and Cottage Grove Avenue (1888 photo). This view shows the distinct forms of the park boulevard and the more common business street. Avery Library, Columbia University.

Chicago's park plans sought to systematize and reinforce trends toward a specialized urban landscape. Similarly, in presenting Chicago park plans, Olmsted and Vaux approved the tendency to "separate domestic more and more distinctly from commercial quarters." They confidently predicted that parks would facilitate that separation by stabilizing neighboring residential areas. In the areas around the south parks no "special inducements to the rise or extension of a commercial quarter" existed, and "the interpolation of the large closed spaces of the Park, turning transportation out of direct channels, will be obstructive to business. . . . The advantage which will come with the Park for securing domestic comfort can hardly fail to soon establish a special reputation for the neighborhood and give assurance of permanence to its character as a superior residence quarter."[101]

Park development also provided a way for the city to compete with contemporary suburban developments. The West park commissioners, for example, considered it a "notorious fact" that wealthy businessmen relied on Chicago's business advantages and yet spent "well-earned" fortunes in the neighboring towns where they resided. Parks offered the "natural advantages" of suburban life without sacrificing the "luxuries and conveniences" of the city.[102] Parks permitted residents to get away without going away. The competition between urban park and suburbs was underscored in Chicago when E. E. Childs, the developer of the model suburb of Riverside, attempted to monopolize the services of Olmsted and Vaux and made "disparaging remarks" about the public park system. Likewise, residents of the suburb of Lake View attempted to defeat assessments for Lincoln Park.[103] The parks and park drives made natural scenery accessible without the "miles of uninteresting travel . . . necessary before the quieter suburbs can be reached."[104] Park proponents appeared less interested in parks as retreats from cities, like the suburb, than with competitively promoting the city as an all-inclusive setting for business, residence, and leisure.

For many Chicagoans, the desire to remove oneself and one's family from the city's commercial downtown undoubtedly involved a desire to move away from the laboring classes of the city. At the same time, some park advocates felt a particular responsibility to provide park grounds accessible to working-class households. By circumstance, rather than desire, these neighborhoods were precisely the ones that suffered most from the indiscriminate mixing of commerce, industry, and residences. Moreover, the public health advantages of urban parks, "the lungs of the city," seemed particularly essential for the city's laborers and their families. In 1869, hoping to influence park legislation, Illinois Health Board member Dr. John H. Rauch published his pamphlet, *Public Parks: Their Effects upon the Moral, Physical, and Sanitary Condition of the Inhabitants of Large Cites.*[105] Rauch recited health and mortality figures in urging the necessity of large public parks. Poorer urban residents who could not travel to or vacation in more healthy areas of the country, he argued, would especially benefit from public parks.[106]

Critics of park development and late-nineteenth-century park reformers challenged the claim that parks served all classes of society. Public grounds like Chicago's South Parks, located five miles from the city center were, according to critics, inaccessible to poor people.[107] For many poor people, working long hours and living at some distance from the parks, these assertions undoubtedly proved true. However, as in other cities, the question of equitable access to parks was both acknowledged and grappled with in Chicago as the major outlying parks were established.

Although the homes of many wealthy Chicago residents stood near Lincoln Park, the area proved easily accessible to poorer neighborhoods along the North Branch of the Chicago River. In 1879 the Lincoln Park Commissioners recognized that many large parks in other cities were frequented "largely by the wealthier class, the visitors in carriages far outnumbering those on foot." However, pointing to the "dense population" residing near the park, the Commissioners maintained that Lincoln Park was "the daily resort of all classes of the community, the poor as well as the rich enjoying the pleasure it affords; the pedestrians far outnumbering those who ride."[108] Similarly, on the west side, the 1877 dedication of Humboldt Park drew a crowd of 20,000, including Irish, German, English, Danish and Swedish immigrants and various national orchestras, bands, and singing societies. Speakers delivered dedication orations in German, English, and Swedish. The dedication ceremonies and subsequent working-class settlement of the west side demonstrated broad use of the parks beyond the carriage trade.

In the late-nineteenth and early-twentieth centuries Chicago's three park commissions built a series of small, neighborhood parks. These parks were designed with playgrounds, field houses, outdoor swimming pools, gymnasiums and modest landscaped areas. They were considered by park commissioners and by later historians as serving working-class residents and as necessary complements of the larger, less accessible parks laid out earlier. Historian Mel Scott writes that Chicagoans were at first lulled into "inactivity" by the gains of the large park movement but were later "awakened" to "the reform efforts to establish small parks and playgrounds in immigrant neighborhoods and other congested areas."[109] This interpretation overlooks the fact that the first parks in most cities were small parks, centrally located, and that in Chicago and elsewhere a commitment to accessible parks led city officials to continue developing small parks even as larger parks arose in the 1860s and 1870s.[110] When, for example, Chicago's Board of Public Works relinquished control of Lincoln Park in 1873 it redoubled its efforts to improve Union, Lake, Dearborn, Jefferson, Vernon, Wicker, Washington, and Ellis parks. The Board undertook ambitious planting and improvement schemes for these smaller, more accessible parks which ranged in size from 2 to 42 acres. The Board of Public Works thought "the smaller parks . . . entitled to consideration" since they served "as resorts of rest and recreation to the working people whose time and means do not permit them to seek the more extensive parks in the suburbs."[111] Having such small sites in comparison with the large landscape parks meant that in the Board controlled parks fountains, pavilions, and bridges often dominated the natural landscape features (fig. 14).

The fight to preserve Chicago's downtown lakefront revealed the park promoters' effort to solve problems of access for poorer residents. In 1869, claiming that the old public ground along the Lake Michigan shore was "almost useless as a pleasure ground,"[112] the Illinois Legislature ordered the sale of the land to the Illinois Central, Michigan Central, and Chicago, Burlington & Quincy railroads for construction of a union terminal. Park lands in other parts of the city, outlined in the 1869 park bills, might be paid for in part with the $800,000 realized from the sale. Strenuous objections arose to a provision granting riparian development rights to the Illinois Central for port and wharf facilities. Despite eventual repeal of this bill, the matter was not finally settled until 1892 when the U.S. Supreme Court ruled against the Illinois Central claim to land and water development rights.[113] Opponents in the battle with the legislature and the railroads resented the commercial "clamoring for the whole lake front." In unusually strong language, a public resolution on the matter, proposed by William Bross and others, declared the park's importance to the health and welfare of all people, "but more especially the poor and indigent among the laboring people, without surplus means to seek outside parks or country

14. Jefferson Park fountain and pavilion (photo ca. 1874–79). Chicago Historical Society, ICHi 22325.

15. Looking south along Lakefront Park and Michigan Avenue (1888 photo). During the 1870s and early 1880s, Lakefront Park was expanded by filling the shallow waters of the lake; in the center the Auditorium Building is under construction, to the right is the Studebaker Building and Burnham and Root's Art Institute with its gable roof. Avery Library, Columbia University.

air. This class have hitherto found their only relief from the unhealthy heats and narrow habitations in the invigorating breezes of our interior sea. The ruthless sacrifice of these classes at the merciless demands of monopoly or wealth, would be a reproach to the humanity and civilization of the State and city in which we live."[114] In the lakefront controversy, wealthy advocates demonstrated some rhetorical and practical concern over physical accessibility for poorer residents during the early period of the park movement (fig. 15).

Indeed, powerful strains integral to the park movement contradicted the desire for exclusive residential areas that mobilized some park supporters. For many Chicagoans, the promise of the "cultivated" ideal lay in its ability to shape a realm of harmonious leisure, uniting city residents across lines of class. In the eighteenth century, American gentlepeople had sought gentility as a means of establishing their own identity as a class; they were largely unambivalent about setting themselves apart from their social inferiors. By contrast, it was characteristic of the nineteenth-century middle class to seek to disseminate its ideals, to enlist others in its own worldview and reform them into its own way of life. The park movement, then, displayed a certain ambiguity. It was the effect of different impulses—one self-involved, the other more broadly social; the desire on the one hand to strengthen, on the other to cross, lines of social class.

As in earlier public grounds and promenades, then, in postbellum parks the dedication of urban space as a resort for all people seemed particularly important. Certainly no other part of the nineteenth-century cityscape counted common endeavor as such an essential part of its social justification. The city street, of course, remained open and accessible to most. However, if the street united people, it did so largely in the pursuit of private interests; distinctions of place and class were manifest. Indeed, the limitations of the street as a realm of acceptable social interaction were apparent in the relative absence of those social actors most essential to cultivated society: middle-class women. The presence of respectable women in parks and on the promenade marked and transformed that space. In parks the hurry, bustle, and anonymity of the crowded city street were supposed to give way to a common pursuit of leisure.[115]

From the point of view of park promoters, the "common" leisure enjoyed in parks nonetheless varied from class to class. While well-to-do visitors were to be saved from the debasing influences of commercial occupations, poorer visitors were to be saved from the debasing influences of commercialized amusements.[116] In a form of positive environmentalism, park advocates suggested that parks provided a more effective means of countering the evils of drinking, gambling, and prostitution than restrictive legislation.[117] Arguing that the "whole difference between virtue and vice lies in the direction in which amusement is sought," the *Chicago Tribune* editorialized that parks constituted a new healthy "direction" for amusements. Promenading seemed a particularly important part of that new direction. William Cullen Bryant applauded this "public and innocent" recreation and contrasted it with the gambling, drinking, and prostitution which took place in "obscure corners" of American towns.[118] In reviewing the commercialized vices located primarily in the city's poorer neighborhoods, the *Tribune* asserted "It is vain to denounce men and women for seeking the lower, more dissipating and sexual indulgences, which destroy the intellect, corrupt the heart and enfeeble the body, unless ample opportunity for other and better amusements are provided."[119]

Incorporating promenade grounds into their park designs, park advocates encouraged the working class to seek leisure along with their social betters. However, in planning large parks the city building elite made no special effort to adjust their designs to accommodate working-class patterns of recreation and leisure. Chicagoans of the laboring classes may have enjoyed the city parks accessible to their neighborhoods, but it is not clear that many sought cultivation on these public grounds. Park promoters offered leisure on upper- and middle-class terms. Similarly, the small park movement at the turn of the century, with its supervised play for working-class children, continued to emphasize middle-class ideals of order rather than working-class patterns of play.[120]

Chicago's park system provided new landscape form to the older ideal which associated horticulture with social cultivation. Like the refined gentleman's private garden, parks pointed to commitments beyond the pursuit of Mammon. The parks created leisure settings closely associated with a domestic realm in which the "home," a new ideological construct, was cast as society's primary agent of cultivation. The wealthy residential sections that parks engendered created leisure and domestic settings pervaded by a sense of refined cultivation. Detached both physically and socially from scenes of commerce and industry, these areas were evidence of wealth and a refined social order. The search for social harmony and efforts to obscure class distinctions in the park movement also gestured toward a social world beyond the city's commercial marketplace.

Chapter 2

"A Different Style of Beauty"
Park Designs, 1865–1880

16. "Under the trees, Lincoln Park" (1872–77 photo). Chicago
Historical Society, ICHi 22324.

Political debates surrounding Chicago's 1869 park legislation contained few clues concerning landscape design. Legislators and voters left the question of what the 1,800-acre park system would look like to the city's park commissioners and their professional advisers. Large tracts of undeveloped landscape designated as public lands but simply left unimproved could have provided many of the experiences called for by contemporary park rationale. Such an approach would answer public health interest in having parks serve as the lungs of the city. It would reverently preserve spots of nature to contrast sharply with the expanding city's built landscape. These areas might even have offered the psychological benefits attributed to contact with nature and removal from artificial scenes of commerce.

Yet park commissioners did not set out to conserve Chicago's natural landscape. In part this was because many people considered that landscape uninviting. Wealthy residents who occasionally expressed concern over artificial commercial life frequently expressed unequivocal dissatisfaction over the city's endowment of natural beauty. In Chicago's "park improvements," the word improvement loomed large. All park designers in Chicago encountered the same difficulty; attempting to create "an antithesis to bustling, paved, rectangular, walled-in streets," they found that Chicago's natural scenes offered little in the way of "grace," beauty, or evocative interest.[1] In 1869, H. W. S. Cleveland despaired of the land designated for Chicago parks: "By what means is it possible to give to areas so utterly devoid of character an expression of natural beauty, and secure enough variety to relieve their monotony?" he wondered.[2] Frederick Law Olmsted, who had encouraged park development for Chicago in the early and mid-1860s, similarly lacked enthusiasm for the "low, flat, miry, and forlorn character" of the city's landscape, which he thought worked against the combination of "urban and rural advantages" in a single setting.[3]

Landscape designers in Chicago thus struggled to reconcile their tasks with a landscape tradition evolved in more promising regions. Their critical assessment of the area's natural beauty did not reflect simple artistic prejudice. If, guided by a highly cultivated aesthetic sense, they found Chicago disappointing, so did visitors and residents who lacked professional landscape training. Although many were impressed by Lake Michigan's grandeur, few visitors at midcentury seem to have found much endearing about the rest of Chicago's natural scenery and topography. In 1849, for example, a visitor who saw great commercial promise in Chicago wrote in his

journal: "The site of the city is such as always to prevent its being a handsome one. It is upon a perfectly level plain, without a hill or mountain to relieve the eye. Upon one side rests the beautiful Lake Michigan, presenting a most lovely water view—yet its cold breezes are any thing but inviting."[4] The designs for Chicago's major parks, then, made during the 1860s and 1870s, all aimed at creating natural beauty from a fairly paltry endowment of natural scenery (fig. 17). In Chicago, as Cleveland put it, "everything must be created."[5]

Yet if Chicago was particularly unlucky, in other cities and other regions park designers similarly intervened extensively to modify and improve upon the given topography. However enthusiastic nineteenth-century city builders waxed about nature, in their parks they sought to have carefully constructed aspects of it. While parks supposedly preserved nature from urban development and business pursuits, park designs shaped nature much like a commodity, created at great expense for a refined urban society. Natural beauty, artistically conceived and artificially created in parks, complemented a wide variety of physical and institutional improvements planned by civic leaders to make cities more desirable places of residence. As elsewhere, in Chicago designers exercised artistic ingenuity in giving park landscape a "nature" that had never existed in the region.

Moreover, love of nature, whether indigenous or artificially constructed, remained only one aspect of the complex, contemporary park ideal. Obvious tensions existed between the solitary contemplation of natural scenery and the gregarious engagements of the crowd (fig. 18). To foster harmonious class relations, to gather people in common leisure pursuits apart from commerce, involved the conscious creation of particular park spaces and forms. Specialized and relatively formal settings such as promenade grounds and carriage drives were needed to accommodate gregarious, ordered interactions among groups of park visitors. So the formal designs for Chicago's parks drew upon earlier forms of the private garden, the promenade, and the public ground.

This chapter looks more closely at the aesthetic of park promoters as it shaped public parks for the city of Chicago. It looks in particular at the ways in which park planners and designers ascribed to particular design principles and design elements the capacity to affect the character of individuals who experienced the landscape and the social relations of groups within the landscape. The idea that landscape might shape thinking, sensibility, and social experience was central to the park rationale. Given the pervasiveness of these assumptions, specific

17. Horticultural display at the Inter-State Industrial Building (photo ca. 1873–92). Standing on the lakefront between 1873 and 1892, the Inter-State Building provided space primarily for annual commercial and industrial expositions. The horticultural display adopted the parks' rustic features and promenade patterns and played a role in the broader promotion of Chicago, its commerce, industry, and culture. Courtesy of the New-York Historical Society, New York City.

18. Lake Shore Drive and Promenade in Lincoln Park (1905 photo). Library of Congress.

design decisions made for Chicago parks are understood as reflections of ideas about the relation between landscape and mentality and the relation between landscape and social relations. The rambles, greenswards, rustic bridges, meadows, and other elements created in Chicago parks held significance in the eyes of nineteenth-century park promoters and urban theorists. In turn, the choices that city builders made in shaping park grounds grew out of their experience of and aspirations for city life. With parks, some Chicagoans attempted to construct an ordered realm counterpoised against the chaotic and conflictual world of downtown commerce. Constructing parks was one aspect of constructing a concept and realm of leisure. Here, park builders sought to manipulate nature and human nature alike; valuing their own reflective and gregarious impulses, park advocates set these qualities against the self-seeking that threatened to dominate their lives. The parks themselves reflected the varied interests that supported park projects and the varied priorities of Chicagoans preoccupied both with sensibility and social relations.

The South Parks

The Olmsted, Vaux and Company plan for Chicago's South Parks represented the most comprehensive plan presented for the city's parks in the 1860s and 1870s. The plan covered improvements for the 593-acre Jackson Park along the Lake Michigan shore between 56th and 67th streets; the 372-acre Washington Park located west of Cottage Grove Avenue between 51st and 60th streets; and the 600-foot wide, 90-acre Midway Plaisance connecting the two parks along the line of 59th Street (fig. 19). The enabling legislation defined the park boundaries. Other than the placement of a park section on the lake shore, scenic considerations did not influence the design as strongly as they did in other cities where the contours of the land often influenced the selection of park sites.[6]

In May 1869, while working for the suburban Riverside Improvement Company, Olmsted toured the South Park area and informally advised the South Park Commission on park designs. Roughly a year later the commission hired Olmsted and his partner Calvert Vaux to draw up a plan for the improvement of the South Parks and their connecting boulevards.[7] In arriving at a design for Chicago's flat topography, Olmsted and Vaux reversed their usual efforts to open quiet, pastoral vistas in urban settings. Working on more varied topography, as in New York's Central Park, Olmsted and Vaux undertook at

great expense to introduce areas of "prairie-like simplicity" to a rougher natural landscape.[8] The restorative value of rural or park scenery derived, according to Olmsted, not so much from picturesque effects nor from overly refined, gardenlike cultivation but rather from simple visual unity and a subordination of detail to general effect.[9] Impressions of great openness, contrasted with the "walled-in" city, became a standard feature of Olmsted design—one enhanced by indistinct planting at the periphery of broad open lawns.[10] Parks that offered these wide horizons might allow visitors to escape the cramped spaces and thinking often enforced in their working lives.

In Chicago, however, broad expanses and limitless views of the horizon proved a drawback rather than an advantage. Nevertheless, Olmsted's general aesthetic preference for scenes of pastoral quiet over picturesque settings meant that he found some promise in Chicago's park sites. The designers began with open spaces and then moderately added the picturesque. "Massive bodies of foliage" and lagoons and lakes that would reflect the foliage provided intricacy and picturesque variety—elements often tamed in other Olmsted designs.[11]

Olmsted and Vaux used Chicago's first prairie to greatest advantage in Washington Park's one-hundred-acre greensward, called the South Open Green. The greensward was surrounded by a grove of trees, walks, and carriage drives. In connection with smaller lawns, the vista across the South Open Green stretched nearly one mile. Jackson Park also had smaller open greens; however, in Jackson Park the natural vistas focused primarily upon the lake, with carriage roads and terraces placed along the shore (fig. 20). Extensive systems of planting around the Jackson Park lagoons and in the south end of Washington Park provided more intricate and diverse views of natural foliage. In Washington Park, Olmsted and Vaux planned a small ramble where deep excavations of paths, richly planted small ledges, and large shrubs would provide a "sequestered and picturesque" character. In balancing elements of the *picturesque* and the *beautiful,* Olmsted revealed his strong commitment to late eighteenth-century English landscape gardening and theory, exemplified in the work of William Gilpin, Uvedale Price, and Humphrey Repton.[12]

Olmsted and Vaux shared the view of Chicago real estate interests that the city's residential area would expand to the outlying parks. They designed the South Parks accordingly. Their design reconciled the two broad types of public parks—the country park or "great roaming grounds," which required a full-day excursion to visit, and the more "garden-like enclosure" used daily

19. Olmsted, Vaux and Company plan of the South Parks, Chicago, 1871 map. From *Chicago South Park Commission Reports* (1869–71), Chicago Historical Society.

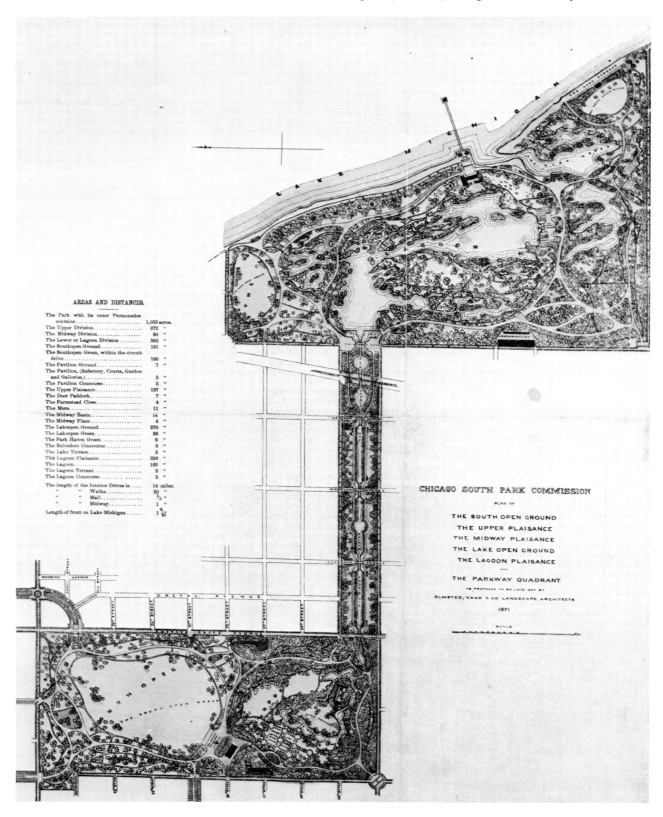

20. Jackson Park's shore drive and promenade (1890 photo).
Avery Library, Columbia University.

and intensively by large numbers of people. They attempted to merge, without undue jarring, the "scale of scenery" of a country park and the "scale of public accommodations" of more active urban public grounds.[13] In 1866, Olmsted and Vaux acknowledged that natural scenery and large numbers of people were "not quite compatible one with the other." After all, the most effective way for a park's natural scenery to contrast with the surrounding town would be to exclude not only scenes and buildings of commerce but the masses of people themselves. "Yet," Olmsted and Vaux reflected, "in a park, the largest provision is required for the human presence. Men must come together, in carriages, on horseback and on foot, and the concourse of animated life which will thus be formed, must in itself be made, if possible, an attractive and diverting spectacle."[14] This clashing of interests expressed itself in designs that provided a variety of distinct park landscapes.

To accommodate the "concourse of animated life" within parks, Olmsted and Vaux incorporated into their designs the older and popular promenade tradition. Their 1858 plan for New York's Central Park included a "grand promenade," which was an "essential feature of a metropolitan park."[15] In park after park, Olmsted and Vaux designed formal promenades, compromising their stated goal of achieving the antithesis of urban bustle. Their designs did not disembody urban aggregation; rather, they attracted large crowds of urban residents for shared leisure, amusement, refinement, and cultivation. Nature, of course, provided part of the attraction for the urban crowd; however, the absence of the scenes of commercial pursuits, which accompanied most other gatherings of the urban crowd, proved an alluring part of the park setting.

In their design for the South Parks, Olmsted and Vaux included spaces that would draw crowds of people together and meet public interest in promenading and in seeing and being seen. In these spaces the pervasive informality of the overall park plan often gave way to a formal design order. In Washington Park, a mall nearly a quarter of a mile long was flanked by straight lines of trees. At the middle of the mall the formal planting set off four square, open-air *apartments* for picnics and gatherings of various associations. At the south end of the South Open Green, a bandstand built in connection with a refectory, a carriage concourse, and vine-covered trelliswork—all formally disposed—provided for promenades, concerts, and large crowds of people. At several points along the carriage roads in the parks, the way broadened to facilitate congregation and conviviality. In

Jackson Park, a terrace along the lake shore permitted carriage riders and pedestrians to promenade along adjacent routes. A lake pier, planned to extend a thousand feet into the lake, was provided for "large assemblies" of people. Olmsted and Vaux anticipated weekly band concerts played on an island in the Jackson Park lagoon. From the nearby refectory lawn, carriage concourse, and formally planted concert terrace, park visitors could easily hear the music performed in the island bandstand. In addition, Olmsted and Vaux designed specific areas of the park for sports—baseball, football, cricket, and running games. They also provided for boating, swimming, and skating areas.

The formal lines and spaces in the South Parks had both practical and symbolic importance. According to Olmsted and Vaux, they facilitated the gathering and movement of large crowds; they clearly oriented and directed crowds so as to avoid confusion or conflicting movement. At the same time, formally designed areas embodied the cultivated possibilities of urban life. Indeed, to the extent that they represented the populated *city,* promenade grounds, concert terraces, malls, and other sites for public gathering made other areas in the park—points of informal, curvilinear, and picturesque design—became more plausible representations of *nature.* In most parks, the space devoted to scenes of nature exceeded the space given over to formal gathering. In this manner a park might strongly suggest a different way of ordering natural and artificial forms in the urban landscape. Within the boundaries of a park, natural form might dominate urban geometry rather than the reverse. Since, in Chicago parks, *"everything"* was created, the parks might represent an idealized urban environment. The promenade grounds and other formal areas for popular gatherings might represent Chicago itself: here, after all, was "urbs in horto," an idealized, commerce-free city located in artificial, tamed, refined nature.

In 1852, Andrew Jackson Downing declared that all artists and "men of taste" agreed that true beauty resided in curved lines, and "the more gentle and gradual the curves, or rather, the further they are removed from those hard and forcible lines which denote violence, the more beautiful are they."[16] Olmsted and Vaux adapted this English landscape aesthetic to their work in an explicitly urban context. In 1868, explaining their proposal for a curvilinear street plan for suburban Riverside, Olmsted and Vaux asserted the desirability of achieving "the greatest possible contrast" between the Chicago street grid and the Riverside layout. The grid signified an "eagerness to press forward, without looking to the right hand or the left." For Riverside, then, the designers recommended gracefully curved lines and open spaces as a means to achieve "happy tranquility."[17] In parks, too, the curve would dominate the line and thereby bring to utilitarian commercial minds a mood of contemplation and appreciation for refined cultivation (fig. 21).

Beyond the purposeful contrast of landscape form with urban geometry, Olmsted and Vaux used bodies of water to dramatize the distinction between the sites of park leisure and the scenes of urban commerce. In Chicago, where the city's fortunes were initially based on waterborne trade, many residents viewed Lake Michigan with a certain commercial calculation. But the lake could be abstracted from commercial associations in a way that city land could not. Water, like the idealized views of other forms of nature, seemed to hover with purity above the impress of commerce; it fit well with the broad aims of park design. Indeed, Lake Michigan's waters sparked greater interest in Chicago's natural scenery than any element found on land. The lake, wrote Olmsted and Vaux, represented the "one object of scenery near Chicago of special grandeur or sublimity." The lake's beauty could not, by artificial means, be improved. However, with consummate artistry, Olmsted and Vaux proposed harnessing the grandeur of the lake and running it, with its powerful associations, through the less scenic park lands. Water would provide the framework for the South Parks plan. Lacing the parks' three divisions into "one obvious system," creating the "impression of unity between Park and the Lake," water would flow between the lake and the lagoons in Jackson Park, and through a canal along the Midway to the lake or mere in Washington Park. These water bodies would receive extensive plantings and allow the introduction of birds from around the world as well as provide an ornithology exhibit. Scenic wonder would eddy along the water route through the park.

As the Lake Michigan waters became an integral element in the South Parks design, the boundaries of the park expanded. Adapting arguments that favored boulevards to connect the city and the parks, Olmsted and Vaux argued that steamboat travel to the South Parks would extend the benefits of the park to the center of the city. The water passage would give Chicago incomparable advantages over parks with more attractive natural landscapes. Ending "uninteresting and tedious" travel over land, the water passage by steamboat would fulfill the cherished vision of bringing the park to the very center of the commercial city: "The park would practically begin at the mouth of the Chicago River," and Chicagoans could enter the park from the heart of the downtown business district.[18] Like curvilinear roads,

21. Washington Park driveways and paths (photo ca. 1870–75). This early view of Washington Park shows the sweeping curves of the carriage roads and pedestrian paths and the planting carried out by H. W. S. Cleveland following the Olmsted and Vaux plan. Chicago Historical Society, ICHi 22320.

waterways would provide a sense of indirection, subtlety, and leisure; they fostered a sense of time and motion that contrasted dramatically with the experience of the city's street grid.

In the end fire took precedence over water in the early development of the South Parks. Six months after Olmsted and Vaux presented their report to the South Park Commission, the 1871 fire swept through the downtown area and much of the built-up portion of the city. It destroyed the commission's offices and records, including plans, accounts, and nearly completed assessment rolls. Even with a new assessment, the money required for land purchases and starting improvements proved hard to collect. Delays raised the cost of park lands from $1,865,500 to $2,600,000. The national economic depression of the early 1870s exacerbated the financial strains created by the fire. In 1877, "the great stagnation in all the Industrial Enterprises" led the South Park Commission to reduce its tax levy by one-third, to two hundred thousand a year, barely enough to cover the commission's bonded debt. Under such financial conditions, the commission sacrificed many elements of the Olmsted and Vaux plan.[19]

The South Park Commission's financial difficulties led it to instruct Cleveland, hired in 1872 to oversee the execution of the Olmsted and Vaux plan, to avoid "making extensive alterations of the natural surface." It also eliminated the proposed water route through the Midway.[20] In 1894, Olmsted revived the plan for the Midway canal as part of his redesign for Jackson Park following the Columbian Exposition. Daniel H. Burnham approved and many citizens signed public petitions in support of the idea, but the economic depression of the 1890s proved no more hospitable to the canal plan than had the crisis of the 1870s.[21]

In the 1870s, improvements did not proceed evenly throughout the South Parks. The earliest ones centered on the Washington Park South Open Green and its surrounding groves of trees. The boulevards leading to Washington Park were also among the park commission's first improvements. Jackson Park, with its more swampy and problematic site, remained largely unimproved during much of the decade.

In spite of financial limitations, by the summer of 1873 the park commission did improve the northern part of Washington Park with trees, shrubs, large greenswards, walks, and carriage drives. The area became a popular resort for day and evening outings.[22] True to Olmsted's intent, those approaching the park by carriage along the "arrow-like straightness" of adjacent streets found themselves pleasantly moving along the park's gently winding road.[23] Following the start of weekly band concerts in 1874, the park commission constructed a permanent band pavilion and concourse along with a restaurant built in Gothic style. Substantial picnic areas developed both in Washington and Jackson parks in this same period. Cleveland planted a twelve-acre ramble with vines and shrubs of from one to fifteen feet in height (fig. 22); its labyrinth of paths proved a "perfect fairy land" for children and a joy for those "of more sedate years." One visitor reported finding the "paradise of the park" among the trees, plants, and swans surrounding the completed section of the Washington Park lakes.[24]

In 1872, when Cleveland took on the South Parks landscape architect position, he confided to Olmsted, "My chief anxiety is about trees. The Comm[issioners] are so desirous of making a great show that I am apprehensive they will urge the planting of very large trees from the woods."[25] Hoping to avoid planting mature trees that would not last, Cleveland insisted on his right to accept or reject all trees delivered to the park. However, the park commission's desire for great show took other forms. The commissioners found a botanical house and various forms of flower gardening useful for attracting people to the parks, even though these features were laid out in undeveloped and monotonous topography (fig. 23). In the 1880s and 1890s, under the direction of Washington Park superintendent Fred Kanst, flowers and plants were sculpted into "floral masterpieces" in the form of formal park gates, sundials, portraits of President Grant and Uncle Sam, various vegetable forms, and men riding in a canoe. Some park observers felt that these gardenesque features were "antagonistic to the pleasure to be obtained from natural scenery."[26] The designs clearly compromised Olmsted and Vaux's strong preference for landscapes that masked artificiality. Nevertheless, the park commission found that floral displays gave "pleasure and delight to thousands upon thousands of citizens and strangers."[27] The floral sculptures and gardens gathered a crowd less interested in reverence for nature than in appreciation of the gardeners' artifice (fig. 24).

Lincoln Park

The improvements of Lincoln Park in Chicago's North Division were closely linked with designs for the city's landscaped cemeteries. The history of Lincoln Park underscored the influence of the rural cemetery movement upon American park design.[28] With cemeteries, as with the gentleman's garden and the city's parks, horticulture

22. Washington Park ramble (1888 photo). These vine arbors formed part of the Ramble developed by H. W. S. Cleveland in Washington Park during the 1870s. Avery Library, Columbia University.

23. Jackson Park Conservatory (photo ca. 1900–10). Library of Congress.

24. Washington Park floral sculpture displays. Floral sculptures designed by Washington Park superintendent and gardener Fred Kanst in the 1880s and 1890s. Chicago Historical Society, ICHi 22316.

and landscape improvement were viewed as gestures of refinement and culture. In 1848, for example, Mayor James H. Woodworth asserted that the lack of horticultural adornment in the city cemetery reflected poorly on Chicago. "Nothing exhibits in a community more clearly a high state of civilization and morality than a proper regard for their dead," a regard evidenced by cemetery embellishment.[29] Besides an interest in ending the "violation of sanitary law," cemetery horticulture drew upon the romantic and noncommercial images of natural landscape to foster a reverence of the living for the dead. In the 1850s and 1860s, Chicago cemetery companies moved beyond the traditional, congested, churchyard style burial ground and followed the landscaped cemetery model established in 1831 by Mount Auburn Cemetery in Cambridge, Massachusetts. Chicago's cemeteries, Rose Hill and Calvary, both founded in 1859, and Graceland, founded in 1860, followed the examples of Mt. Auburn, Philadelphia's Laurel Hill, Brooklyn's Greenwood, and Cincinnati's Spring Grove in providing burial sites on large outlying tracts amid extensive horticultural improvements. As popular resorts for urban excursionists and picnickers, these cemeteries educated popular taste in landscape design.[30]

The opening of Graceland, Rose Hill, and Calvary ended burials in Chicago's old city cemetery. The large tract north of the cemetery initially designated for cemetery expansion proved ideal for the development of what became Lincoln Park. In later years, as the park expanded, the old graves were moved to the city's newer cemeteries. The Common Council committee that recommended the establishment of the park at the old cemetery site pointed enthusiastically to the somewhat uneven ground surface and suggested that the stream of water running through the site might be developed as fish ponds or small lakes—good for boating and skating. The report also envisioned the fine views possible over Lake Michigan as central to the park's scenic possibilities.[31]

In 1865, following the decision to name the park in honor of President Lincoln, the Common Council seriously embarked upon park improvements. Swain Nelson, a Swedish landscape gardener, planned the park, and the Board of Public Works began excavating the ground for small artificial lakes (figs. 25, 26). Dirt from the excavation was used to create small hills, and the work of sodding lake shores and grading the ground began. In 1867, Nelson and his partner Olaf Benson also supervised the design of a small lake in Union Park. As annual expenditures rose from $4,500 in 1865 to $32,000 in

1869, Lincoln Park became the showcase park of the city, "thronged with visitors" who came to enjoy the improvements and listen to concert music.[32]

The Lincoln Park bill met legal challenges which caused vexatious delays of the improvements. In 1875, after additional legislative acts and several legal cases, the courts finally upheld the Lincoln Park assessments.[33] Olmsted's friend Ezra B. McCagg, an early Lincoln Park commissioner, stated that good taste suggested a park with slight surface variation, small artificial lakes, comfortable carriage drives, vistas over Lake Michigan, and a recognition that "simplicity is not inconsistent with beauty."[34] In McCagg's view, the design would stand in striking contrast to the complexity and topographical variety of New York's Central Park.

During the 1870s, Nelson continued his work as chief designer, gardener, and contractor in Lincoln Park.[35] In 1879, the commissioners reported that the park included 120 acres of lawn and planting area, $6\frac{3}{4}$ miles of drives, 7 miles of walks, 12 acres of lake, and 75 acres yet to be improved. A continuous drive of $2\frac{1}{4}$ miles running through the park and south to Pine Street provided excellent views out over the lake. The park included many of the attractions found in the South Parks: fountains, a music stand, pavilion, formal mall promenade, greenhouses, some spots of formal walks and gardens, artificial lakes, rustic bridges, separate and distinct carriage and pedestrian paths, trees, flowers, and shrubs—all disposed in a largely informal manner. Whereas Olmsted and Vaux advised against zoological collections in the South Parks, in the 1860s both Lincoln and Union parks displayed animals. More than any other Chicago park, Lincoln Park received, in the 1880s and 1890s, a large collection of memorial and honorary statuary. In 1886, the commission started construction of a 90-acre breakwater and lake shore promenade, built 300–500 feet from the shore of the park. Accessible by bridge, the breakwater promenade created an extensive inlet and lagoon area along the old shore line (fig. 27).[36] Lincoln Park thus combined areas for gregarious and solitary recreation.

The West Parks

The West Side Parks lacked the redeeming grandeur of Lake Michigan. Humboldt, Garfield, and Douglas parks, 200 acres, 185 acres, and 180 acres respectively, also had less space with which to compensate for their natural disadvantages. In an area about the size of Washington Park's South Open Green and adjacent walks, drives, and tree groves, the West Park designers compressed all the

25. Boating in Lincoln Park, Grant Memorial in distance (photo ca. 1900). Chicago Historical Society, ICHi 22304.

26. Skating in Lincoln Park, Grant Memorial in distance (photo ca. 1895). Chicago Historical Society, ICHi 22326.

27. Lincoln Park lagoon and high bridge (1900 photo). Library of Congress.

desirable forms and accessories found in the other parks. As a result, the smaller area affected the aesthetic balance between formal and informal elements in the West Parks designs.

The West Chicago Parks bill did not specify the precise location of the parks. It called for three parks of from 100 to 200 acres located within the city limits, one north of Division Street, one south of Harrison Street, and one between these streets. In the hopes of capitalizing on speculative interest in park location and gaining donations of park land from developers, the commissioners advertised ten possible locations for the parks in July 1869.[37] They also offered to name the parks after a person or persons who donated a large sum toward park improvement. Soon afterward, they added three more sites but resisted suggestions that the middle park be moved closer to downtown, since that would have increased land prices and decreased the size of the park.[38] Selected in November 1869, the final park plan established the east lines of the northern park, Humboldt, and the southern park, Douglas, along California Avenue. The main body of the middle park, Central Park (renamed Garfield Park in 1881 after the assassinated president), was placed a mile west of California Avenue.

As in the case of other city parks, assessments of the artistic possibilities for the West Parks were tempered by the sobering realities of nature. The earliest discussion of design for the parks came in July 1869, when the surveyors Alexander Wolcott and Edward A. Fox completed their topographical survey and report for the West Park Commission. They reported, "It would seem at first glance to be a difficult if not impossible undertaking to transform this flat, treeless, uninviting prairie into a pleasure ground that should possess those attributes of picturesque variety and beauty that we are accustomed to associate with the name of . . . public parks." The area lacked many of the "natural advantages" that made New York's Central Park so attractive. However, in "a different style of beauty," real possibility for improvement existed. Although Wolcott and Fox failed to receive the commission for the final West Parks design, their report outlined the main features of the plan eventually undertaken. They suggested achieving some "diversity of surface," giving an occasional elevation of a few feet to break the monotony of the ground. Artificial lakes and fountains would form "one of the most attractive features." Their plan called for tree, shrub, and flower planting and well-graded carriage drives and footpaths, rustic bridges, and "other tasteful structures."[39]

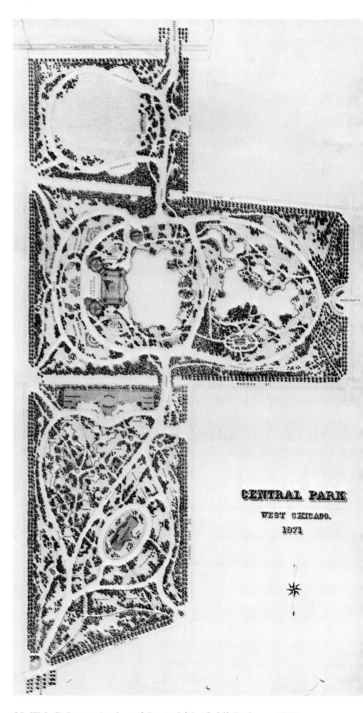

28. W. L. B. Jenney's plan of Central (Garfield) Park, ca. 1871. From *West Chicago Park Reports* (1871), courtesy of Chicago Park District Special Collections.

In late 1869, the West Park Commission informally consulted Wolcott and Fox and Jenney and Schermerhorn regarding designs for part of the West Parks boulevard system.[40] In 1870, the commission hired architect William L. B. Jenney and his engineer-partner to design the entire West Parks and boulevard system. Jenney's temperament and cultural outlook coincided with that of many park proponents in Chicago. In 1865, for example, writing to Olmsted in the hope of obtaining work in New York, Jenney expressed his desire to move from his army position in St. Louis to the East; "In the West," he wrote, "there [is] little knowledge and little desire for Art."[41] Thus, when Jenney eventually settled in Chicago, he appeared to be willing to join with civic leaders in raising the level of artistic and cultural life and redeeming the city from the prejudice reflected in his earlier critique. Jenney saw himself as a crusader aiming to "improve the taste for art." He thought that art should grow from and embellish commerce, creating "oases in a desert."[42]

Jenney's plans for the West Parks contained the main features outlined by Wolcott and Fox and shared common stylistic features with Lincoln Park and the South Parks.[43] Curvilinear carriage drives and pedestrian paths, broad greenswards, informal planting, and a dominant body of water were included in all three of the West Parks. Like the Olmsted and Vaux plan for the South Parks, Jenney's landscape design destroyed the grid geometry of the parks' approach roads. In his design for the boulevards running between the parks, Jenney insisted on formal design, rejecting the idea of making these roadways curve from side to side. He did decide, however, that when any of these stately though "somewhat monotonous" boulevards ran through the park, it would "completely change its character." On entering the park, formality would give way to informality.

In the case of Garfield Park, city officials insisted that Madison and Lake streets run a straight course through the public ground. Jenney considered the formality of these streets and their anomalous scenes of commerce and traffic antagonistic to park design and use. He proposed densely planted embankments to hide the streets from the view of people in the park and linked the three divisions of the park with viaducts (fig. 28). His strategy of removing park visitors from scenes of commerce on the streets revealed central elements of his park philosophy. The primary object of the park was to "give relief for a while from all intercourse with the toil and humdrum of life" and to help the businessman "forget the anxieties of the counting house" and the laborer and artisan to "forget his toil."[44]

In spite of Jenney's interest in excluding scenes of commerce and artificiality from the parks, his design did include major points of formal, axially conceived order. He clearly drew heavily upon the architecture and engineering tradition of his French technical education. His designs were also based upon the realities of the constricted sites with which he worked. His commitment to including the popular and formally designed areas for the gregarious gatherings held in Chicago's larger parks meant that formal spaces stood out more prominently in the smaller West Parks. In sharp contrast to Olmsted and Vaux's plans, Jenney created formal terminal points at the intersections of the boulevards and parks. In Garfield Park, a Washington memorial column terminated one main approach; in Douglas Park a large formal fountain and memorial gate terminated the Douglas Boulevard connection (fig. 29). A formal boulevard continued along a park drive into the center of Humboldt Park, where it widened to encompass a grand formal plaza with a monument, fountains, and a symmetrical pavilion overlooking a small lake (fig. 30). In Olmsted's parks the formal spaces were generally discontinuous from the geometric order of the boulevards and surrounding street grid. Visitors arrived at these formally designed areas only after traveling some distance through the park. Jenney made the links more continuous and obvious to park visitors approaching the park periphery. He felt that parks, on analogy with buildings, needed properly signified and developed entrances. The West Parks plan included many spots for the gathering of the crowd—boat landings and terraces, music pavilions, refectories, carriage concourses, museums, zoological gardens, botanical gardens, plant and propagating houses, rustic shelters and benches, dairies, public toilets, picnic and bird sanctuary islands, children's playhouses, lawns for croquet, baseball, and military parades.[45]

In the first ten years of park improvement the West Park Commission carried out many of the general features of the Jenney plan. Over ninety-six thousand trees and shrubs were planted and over eight hundred thousand cubic yards of dirt were excavated in building the lakes and grading paths, drives, and lawns. Although the commissioners did not establish museums or zoological collections, they did open plant conservatories and palm houses. Water birds were placed on islands in the parks' lakes. The major improvements focused on these lakes, which were from 11 to 16 acres large. The park refectories, ornamental boat landings, music pavilions, and carriage concourses were all developed in concert along

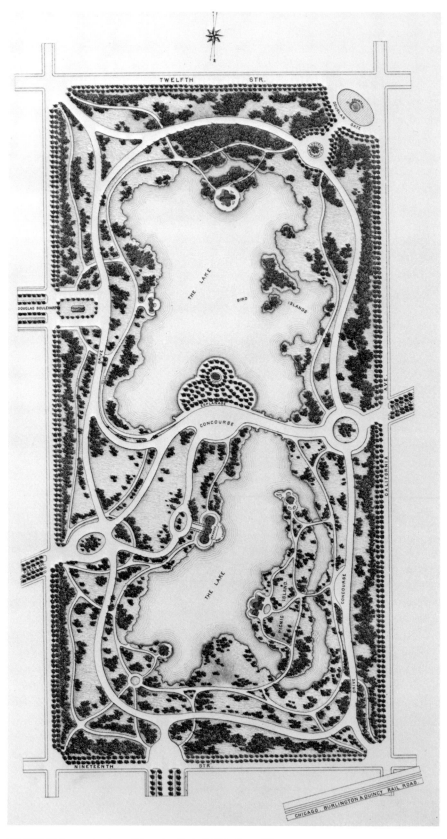

29. W. L. B. Jenney's plan of Douglas Park, 1871. From *West Chicago Park Reports* (1871), courtesy of Chicago Park District Special Collections.

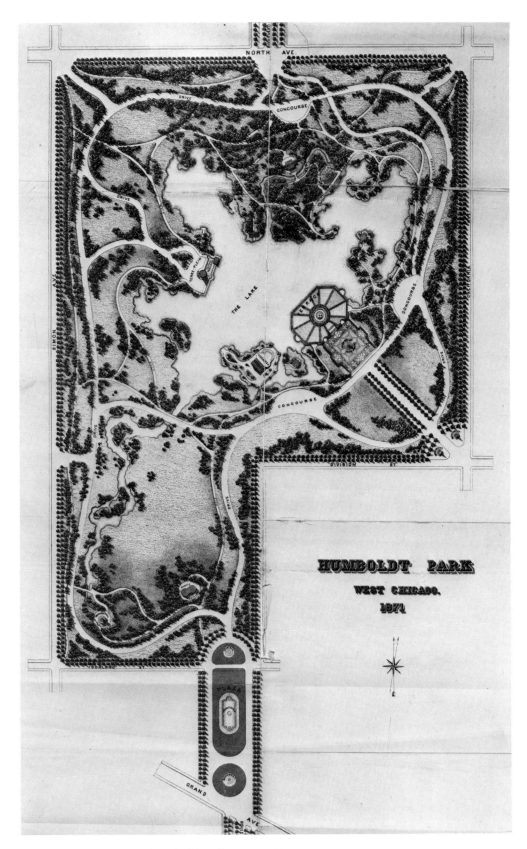

30. W. L. B. Jenney's plan of Humboldt Park, ca. 1871. From
West Chicago Park Reports (1871). General Research Division;
The New York Public Library; Astor, Lenox and Tilden
Foundations.

the shores of the lakes (fig. 31). The popularity of boating provoked constant comment. Just as plantings varied from formal to informal, the park accessories—bridges, benches, fountains, and shelters—ranged from sober, somewhat classical designs, such as the Garfield Park bridge rimmed with ornamental vases, to playful rustic designs for log benches and wood-crafted park shelters.

In the West Parks, as in the parks of the city's other divisions, planners anticipated the coalescing of the urban multitude around shared recreational and leisure activities (fig. 32). Setting aside commerce, at least momentarily, people appeared in the parks as the virtuous followers of the fine arts of horticulture, music, and refined civic pursuits. The promise of the commercial city, conceived in grain elevators, meat packing plants, warehouses, factories, and office buildings, seemingly bore fruit in the "curious fact" of the urban park.[46]

Boulevards

In their accommodation of many aspects of urban life, the city's boulevards and parkways followed the pattern of city parks. On the boulevards, with scenes of the city close at hand, city residents mixed an appreciation of cultivated nature with an enthusiasm for the pageantry, parade, and possibility of the city crowd. In the late 1840s, John S. Wright proposed connecting a series of outlying parks with *parkways* and drives along the lake.[47] In subsequent years nearly every proposal for Chicago parks included plans for a system of boulevards or landscaped avenues. The 1867 South Park bill, passed by the legislature and defeated by the voters, for example, provided for a two-hundred-foot-wide boulevard on line with Elm Street, to run from the park, west of Cottage Grove Avenue, to the lake. All three of the park bills passed in 1869 contained provisions for boulevards. From the outset, the separate park commissions worked to join the boulevards in a single, continuous system of pleasure drives. On the boulevard, park and city merged. In 1872, the *Chicago Tribune* reported that "the boulevards are parks 'spun out' ".[48] Spinning their way through the city, the boulevards joined the parks and the city as the warp and woof of real estate development and creative urbanism.

Boulevards and park drives offered important urban amenities *that* had been proposed earlier and quite apart from the park movement. For example, they provided well-paved roads, a long-sought-after and rarely attained object of municipal improvement. The level of beauty and comfort of well-constructed park boulevards cannot

be underestimated in a city that had been for so long mired in mud and road ruts. Providing the smoothest, finest driving surfaces in Chicago, the boulevards proved tremendously popular. Boulevards also included extensive borders of ornamental shade trees, which many people admired for their beauty and as evidence of a civic appreciation of fine art.[49] Before the development of boulevards, efforts at planting trees along Chicago streets had proved extremely sporadic and inconsistent. Following his move to Chicago in 1869, Cleveland crusaded for both tree planting and boulevard development. He considered systematic tree planting along Chicago streets as the single most powerful means to give the city a "distinct character of refinement and elegance.[50] Cleveland proposed a municipally controlled "*system* of street planting to be pre-arranged by careful designs instead of leaving the work to be done or not by the proprietors of lots in the present hap-hazard manner."[51]

Cleveland's tree planting proposals dramatize the degree to which the city builders' ideal of cultivation and the aesthetics associated with it entailed a neatly ordered and compartmentalized urban landscape. In his vision, beautiful landscaped boulevards would replace the utilitarian grid, yet those boulevards would nonetheless clearly define distinct and bounded districts of the city. To eliminate the "confused mass of foliage" that resulted from uncoordinated tree planting, certain areas would be planted with maples, others with lindens, or tulip trees, or horse chestnuts. Cleveland's plan gave neighborhoods a horticultural distinction and rendered comprehensible a quickly expanding, changing urban landscape.[52] That the control of park boulevards was centralized made them an ideal setting for systematic tree planting. Seeking to offer the same amenities of boulevard residential property, some large Chicago real estate developers, including Samuel J. Walker on Ashland Avenue and Samuel E. Gross on Michigan Avenue, emulated the park commissions' coordinated tree planting.[53]

Beyond smooth driving surfaces and trees, boulevards offered broad lawns and specialized spaces for pedestrians, equestrians, carriages, and wagons. More importantly, boulevards helped forge a broader voter coalition favoring park development. Boulevards, or parks "spun out," spread the benefits of parks to neighborhoods located some distance from the parks themselves. They drew people primarily interested in the extrapark matters of street extension and real estate improvement into the park movement. Boulevards provided greater access to park amenities to people who lacked the time or resources to travel to the larger outlying parks. Boulevards also appealed to those interested in developing a

31. Visitors in Garfield Park (1907 photo). Library of Congress.

32. Bicycle riding in Humboldt Park (1888 photo). Avery
Library, Columbia University.

great city, making it attractive to residents who might otherwise seek suburban residences. They also promised something that other cities, particularly New York, lacked in the way of civilized urban amenities.

The boulevards juxtaposed leisure recreation, natural landscapes, and vivid urban scenes. The familiarity of the city crowd, the spots of formal design, the diverse accommodations and contrasts of the large urban park have often been obscured by the breadth of the natural scenes and the sentimental and often crusading views of their chroniclers. In the boulevard, this dynamic juxtaposition of natural and urban scenes was inescapable. Nature's benefits could be felt even as people hurried along the road on their way between home and business. Cleveland, other designers, and real estate developers expected that chains of boulevards and small parks would attract residences, public buildings, hotels, and "magnificent shops."[54] Cleveland, who supervised the development of the South Park boulevards, said that "their attractive interest lies in the charming effect of the contrast produced by the introduction of the most beautiful forms of vegetable life in the midst of the richest architectural decorations and the busiest scenes of the city." He felt that boulevards surpassed parks in terms of both beauty and utility.[55] In 1873, the *Tribune* proudly editorialized that the "panorama of rare attractiveness" along Chicago's boulevards and parks made boastful New Yorkers and other visitors suddenly aware of how "small, contracted, and limited" Central Park's attractions were.[56]

Cleveland's belief that the concourse of life could prove an attractive feature of the park boulevards derived in part from his distaste for "monotony" and the "dead flat" of Chicago's natural landscape.[57] In analyzing various boulevard projects Cleveland insisted that "the interest of the place must be *intrinsic* instead of extrinsic and that this is to be accomplished by making it throughout a school of the instruction of the people." Sections of the boulevard system could be devoted to different didactic forms, including a nursery displaying arboriculture, experimental agricultural plots, fish ponds showing fish culture; different building materials could be demonstrated on the road surface, and the trees, clustered by species, could make the entire boulevard an extended arboretum. Proposing that Chicago avoid the "vulgar" and hopeless attempt to imitate cities with more beautiful natural scenery, Cleveland wrote that the city's only hope lay "in the great museums—so to speak whose intrinsic interest should absorb the attention."[58] The artificiality of this *museums* approach to natural

scenery paralleled the boulevards' human parade and scenes of city life.

Chicago's boulevards ranged in width from two hundred to four hundred feet and thus gave ample room for varied plantings and other uses. American design generally followed European precedent and provided for extensive plantings, pedestrian walks and for a separation of foot, equestrian, pleasure carriage, and business wagon travel.[59] In more formal plans, lines of trees defined and separated the different modes of traffic. Olmsted, Cleveland, Jenney, and other designers were familiar with Avenue de l'Imparatrice between Bois de Boulogne and Paris, which at the time contained informal planting, lacked lines of trees, and had a wooden fence separating the carriageway from the pedestrian way. In Brussels in the late 1850s, Olmsted noted that lines of trees did separate the various forms of traffic.[60] Olmsted wavered in his earliest American proposals for boulevards between seeking a *picturesque* composition, "more park-like than town-like," and proposing more formal, stately designs.[61] The plans for the South Parks boulevards embodied both approaches to boulevard design.

Olmsted and Vaux planned their first Chicago boulevard for the Riverside Improvement Company. The 150-foot wide-road connecting suburban Riverside with Chicago would, if carried out, have provided an enjoyable route to Riverside, "neither tedious nor fatiguing." The plan aimed at the "gratification of the gregarious inclination," best exemplified by the European promenade tradition—a "social custom of great importance."[62] Olmsted and Vaux reported, "It is an open-air gathering for the purpose of easy, friendly, unceremonious greeting, for the enjoyment of change of scene, of cheerful and exhilarating sights and sounds, of various good cheer, to which the people of a town, of all classes, harmoniously resort on equal terms, as to a common property. There is probably no custom . . . more favorable to a healthy civic pride, civic virtue, and civic prosperity."[63]

In a manner quite similar to Adolphe Alphand's division and planting of Paris boulevards, Olmsted and Vaux recommended a 36-foot-wide carriage road with a border especially adapted to equestrian riding. Flanking the horse trail on either side, and moving toward the outside property lines, was an 8-foot lawn with a line of trees, an 8-foot walk, another 8-foot lawn with a line of trees, a 20-foot road for local delivery wagon traffic, an 8-foot sidewalk, and a 4-foot-wide hedgeline.[64] At intervals along the road, Olmsted and Vaux proposed widening the boulevard to provide special gathering spots designed with

33. Olmsted and Vaux plan for parkway from Chicago to suburban Riverside (1869). Library of Congress.

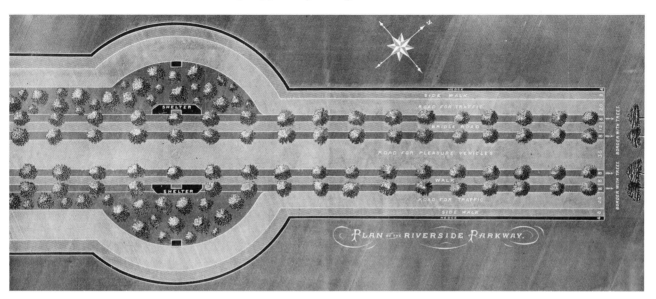

informal plantings, ornamental decorations, watering places, and sheltered seats (fig. 33). The landscaped gathering spots served as points of destination and turning for rides out from the city.[65]

Although financial difficulties led Riverside's developers to give up the boulevard plan, Olmsted and Vaux adapted the plan for two 1½-mile-long boulevards in the South Parks. The 200-foot-wide roadway known variously as Kankakee Boulevard, Grand Boulevard, South Open Parkway, and today, Martin Luther King Drive, included a 55-foot-wide carriage road flanked on either side by planting and walking grounds, local traffic roads, and peripheral sidewalks. The boulevard's formal character and three lines of trees planted on either side of the carriage road drew upon those elements of the Riverside plan directed toward expeditiously moving people (see plate 4). It aimed at "rapid movement of a great number of persons" between the settled neighborhoods to the north and Washington Park.[66] A second boulevard, Drexel Boulevard, also ran north from Washington Park. Its character suggested the informal gathering spots of the Riverside plan. Olmsted and Vaux planned the area for more leisurely movement and for "inviting rest and contemplation."[67] On Drexel, a 100-foot-wide central planting strip separated two 40-foot-wide carriage roads. The central strip was planted with clumps of trees, bushes, and ample flower beds. It had meandering footpaths through the central area. Drexel Boulevard came closest to the model of the boulevard as a small, accessible park area (figs. 14, 34).

Chicago's other park boulevards took their shapes according to a variety of other plans. Garfield Boulevard, 200 feet wide, extended west from Washington Park, 3½ miles to Western Avenue, followed the Drexel Boulevard design, with a 90-foot-wide central planting strip, flanked on either side by two 40-foot-wide driveways and two 15-foot walks. Along the Western Avenue Boulevard, extending north, all the planting area was consolidated to form a more parklike appearance. The West Park boulevards were delayed for years by court litigation over land condemnation. Nearly all 250 feet wide, they eventually adopted a variety of plans similar to the South Park boulevards. Their most distinctive feature was that at points where the boulevards changed direction, the road opened into squares 400 feet by 400 feet, providing both formal points and small public gardens. In Palmer Place, the 400-foot width was extended for 1,700 continuous feet, creating a small park of twelve acres flanked by two roadways. In Logan Square, the area opened was 400 by 700 feet. In Lincoln Park, one of the earliest improvements undertaken in the 1870s was the extension of a boulevard along the lake shore from North Avenue south to Pine Street for about three-quarters of a mile. A carriage drive was lined on either side with trees; a broad mall with plants, walks, and benches was constructed on the shore side of the drive. As in the South and West parks, the Lincoln Park Commission took control of several city streets leading to the park, improving them with good pavements and fine lines of trees (fig. 35).[68]

34. Drexel Boulevard (ca. 1870s). Chicago's boulevards spread
park amenities to residential streets. In contrast to the formal
lines of many Chicago boulevards, Drexel Boulevard reflected
the more informal design of the parks themselves. Chicago Historical Society, ICHi 22301.

35. Washington Boulevard looking east from Garfield Park
(1888 photo). Beyond its specially designed park boulevards,
the park commissions took control of tree planting along some
of the main streets leading to the parks. Avery Library,
Columbia University.

Chicago's boulevards quickly attracted great popular attention and approval. In 1874, Cleveland wrote to Olmsted that he would be amused to "see how delighted our people are with the new toy—the Park & Boulevards—By careful count on a recent evening . . . between the hours of 6 & 8 P.M., no less . . . than 4,600 equipages passed over the Boulevards. In short the jam is such that stringent police regulation is necessary to prevent disorder."[69] The boulevards asserted "a decided influence on Social Chicago." The newspapers reported evening drives on the boulevards much as they reported local charity or society balls—gentlemen and companions were named, equipages and dress described, and fashionable pageantry pictured.[70]

The "great moving masses of fashionable turn-outs," the "motley" assortment of vehicles, the "array of toilettes," and the cosmopolitan mixture of "personages and nobodies," all promenaded on to the boulevards (fig. 36). These contrasts, the sights of both carriages and pedestrians, suggested the realization of park proponents' ideals of class mixture—at least among the relatively prosperous classes.[71] Outlining similarly linked boulevard and promenade arrangements for Boston, one park advocate stated that the "scheme meets the wants of both rich and poor. . . . It adds to the pleasure of those who ride to be surrounded by wandering crowds of pleasure seekers on foot. It increased the enjoyment of those who walk to watch the elegant equipages of those who ride." Boulevard designs distinguished between those who walked and those who rode in carriages or on horseback, laying out specialized spaces for different classes of people and activities. Along Chicago's Grand and Drexel boulevards, one observer noted the pattern of social specialization: "the richly attired ladies and stylish looking gentlemen reigned supreme, and the common people did not block their way."[72] In separate but mutually visible drives and walkways, the social classes might come together without mixing indiscriminately.

In June 1876, the South Park Commission began running public phaetons over the boulevards for those who did not own carriages. In the first 117 days of operation, 24,733 passengers paid ten cents each to ride the phaetons along the South Park boulevard circuit.[73] In 1873, the South Park and Lincoln Park commissioners attempted to make the parks and drives even more popular by suspending the eight-mile-per-hour speed limit on two evenings each week. Although they forbade "matches or boisterous conduct," the park commissioners showed themselves amenable to allowing fast driving and other activities that would furnish amusements quite

apart from the quiet contemplations of "rural simplicity."[74] The parks and boulevards provided a new congenial landscape for urban residents committed to the idea that "human nature is gregarious."[75] They also added to the "many very powerful attractions in a great city to men and women of refined culture which cannot be had in the country."[76]

Yet in spite of their cosmopolitan variety, the parks and boulevards excluded many popular commercial recreations such as gambling, alcohol consumption, and vaudeville performances.[77] These activities, found in abundance throughout the city, were viewed by the elite as debasing to character and thus inappropriate in the park. The South Park Commission forbade "threatening, abusive, insulting or indecent language" and directed park police to evict all "drunken, disorderly, or improper persons, and all persons doing any act injurious" to the Park. In keeping with traditional attempts to provide public grounds and promenades somewhat removed from commercial associations, the commission excluded from the park roads and carriageways of the boulevards all vehicles transporting commercial goods, merchandise, building materials, manure, soil, and other articles.[78] With the baser aspects of commercial life relegated to other parts of the landscape, park and boulevard visitors could, in the view of park proponents, coalesce around shared leisure pursuits.

In 1893 one visitor to Chicago's parks cheerfully declared, "Nature has been tamed and civilized and her ruggedness and her softness woven into a garment for the earth."[79] Just as parks tempered an ungainly natural landscape (fig. 37), they were supposed to temper an objectionable cultural landscape—one pervaded by Mammonism and hard commercial calculation. In fact, the rhetoric of retreat found in park discourse could be exploited to promote the city's growth. The *Chicago Tribune*, for example, viewed parks not as a way of getting away from the world but rather as a way of getting the world to come to Chicago. Parks combined with fine hotels, theaters, churches, museums, "metropolitan amusements," refined society, and good weather to recommend the city as an important summer resort (fig. 38). Reflecting the unsentimental view of nature of many park advocates, the *Tribune* reported that anyone familiar with the country who visited the parks could

> scarcely deny that Nature commends itself more with modern improvements than it does without them. . . . The advantages of improved Nature are obvious. A Saturday in Lincoln Park alone is sufficient to establish its superiority over wild and uncultivated Nature. You

36. Grand Boulevard looking south from 35th Street (photo ca. 1900). Chicago Historical Society, ICHi 20804.

37. Lincoln Park conservatory and flower beds (1900 photo).
Library of Congress.

38. Jackson Park lagoon and the Gallery of Fine Arts built for the World's Columbian Exposition (photo ca. 1900). One of the few buildings left from the Exposition, the structure housed the Field natural history collection until a building for it was completed on the downtown lakefront. Library of Congress.

have hills which are not too high to climb, sheets of water whereon your boats may float with the security, if not the grace, of a swan; banks to recline on without being haunted by the terror of snakes or black bugs; good roads to drive on that are never dusty or too muddy; rustic bridges to cross without the danger of breaking through; . . . pretty women, equipages, and the greeting of friends to enliven the scene, picturesque views that are panoramic in the constant changes the people make; the delusion of bucolic pleasures without the stern reality of bucolic fatigue, monotony, and stupidity.

You could have all this and still be within easy reach of a "comfortable city home," complete with a bath, insect screens, and comfort all around you.[80] In civilizing the city with parks, city builders abstracted nature to suggest a cultivated urbane sensibility.

The gregarious human presence and scenes of *improved nature* encountered in urban parks and along the boulevards did not appeal to everybody. In spite of his interest in juxtaposing natural landscape and urban buildings along the boulevards, Cleveland took strong exception to the low level of public interest in nature. Pained by the intrusions on the "pristine wild beauty" of many country areas and the superficial designs of nature in the city, Cleveland wrote to Olmsted, "I get heart-sick and disgusted at times with the twaddle that passes for 'love of nature'—in the face of the evidence we have everywhere that to the great mass of the so called cultivated people—nature has no attraction except when aided by the merest clap traps of fashionable entertainment which the real lover of nature seeks to escape from."[81] Cleveland suggested that the need to develop "two systems of parks—one for the comparatively few who really want seclusion and the beauty of nature—another for the multitude who can only enjoy solitude in crowds—to whom any work is artistic in proportion as it is artificial—and who unfortunately exert the controlling influence in almost every community."[82] Chicago's park designs, and especially those for the South Parks, attempted to reconcile this tension between solitary nature and the festive gathering of the crowd. Rambles and open grounds at certain hours offered solitude; pavilions, boulevards, promenade terraces provided a pleasant setting for the crowd. The divergence between the solitary and crowd constituencies of the park and the artifice of *natural* design and appreciation points to the divergent interests at work in the nineteenth-century park movement.

Chapter **3**

"A Parallel Moral Power"
Churches, 1830–1895

39. First Baptist Church, Wabash Avenue looking north from
Peck Court, W. W. Boyington, architect, 1866 (1866–71
photo). View shows the First Baptist, Wabash Avenue Methodist Episcopal, and First Presbyterian churches and, barely
visible behind the First Presbyterian tower, St. Paul's steeple.
Chicago Historical Society, ICHi 22335.

Chicagoans viewed the religious landscape as a symbolic commentary on their culture, and church building proved central to demonstrating the city's religious and moral commitments. As Rev. John P. Gulliver put it in 1865, strangers to the city were "profoundly impressed with the material power" of Chicago. In viewing the city, they faced the anxious question of "whether there was here a parallel moral power" that might temper the effects of Mammon. The cornerstone laying for the city's New England Church led Gulliver to optimism on the subject: "We look at you as standing in the great Thermopylae of moral conflict," he told gathered church members, "and when we see you exhibiting an enterprise and industry in spiritual matters parallel to the business energy of your citizens . . . we look on with intense interest."[1] Like many others, then, Gulliver was apt to use a material rather than an ethical standard for gauging the city's "parallel moral power." Grand churches, like large landscaped parks, involved the commitment of material resources to seemingly nonpecuniary or at least nonproductive activities; they could be read as evidence of the transcendence of narrow self-interest and signs of civilized attainment.

Chicagoans inherited a tradition of thought that opposed the realms of God and Mammon, heaven and earth, religious and secular affairs. The segmentation of the city's neighborhoods and the domination of downtown by commercial building forms posed a dilemma for those who shaped Chicago's religious landscape. In 1870, echoing assertions that pervaded local cultural debates, Chicago's Presbyterian weekly declared, "commerce and industry should grow . . . but they should be consecrated to the higher and better uses of life—should be made a means, not an end. Mental culture and Christian living should go hand in hand with business culture, and not be sacrificed to it."[2] For revealed religion to have its desired effect, religious architecture could not remain static in the face of growing commercial monumentality. This chapter looks at the ways in which Chicagoans created religious buildings to represent (and perhaps support) their aspirations for a moral self and a cultivated society. It suggests, then, that their attitudes toward commerce and class significantly shaped the city's religious landscape.

The single most notable development of Chicago's nineteenth-century religious landscape related not to church form or style but rather to church location. Starting in the 1850s, one congregation after another abandoned the center of the city. The 1871 Chicago fire dramatically concluded a process that was already well

advanced: few churches provided a counterpoint to commercial buildings in the downtown, and after the fire only a handful were rebuilt in the area. Instead, church buildings were moved to residential sections where, through their size and style, they visually dominated their neighborhood.

This form came to characterize the modern American city. Yet in the early nineteenth century, a churchless city center was an anomaly. Following patterns of European urbanism, the seventeenth- and eighteenth-century builders of American towns generally located houses of worship at or close to the center of towns. When Francis Nicholson, a colonial governor, planned Annapolis and Williamsburg, he placed the Anglican church on prominent central sites. In Annapolis, capital city of the colony of Maryland, the street system radiated from the *state circle* and from the *church circle,* each located atop the promontory that overlooked the town and the harbor. Commerce and residence filled the interstices and were formally overseen by church and state. In the many places where towns were planned less formally, settlers frequently located their houses of worship in central sites that related closely to public lands or buildings. Putting meeting houses in the center often expressed the assumption that community and congregation would coincide. Even Puritan New Englanders, who separated church membership from community membership, nonetheless expected everyone in the town to attend Sunday service. Moreover, New England meeting houses served for town meetings and other public gatherings. In this context, it was important that meeting houses be located centrally to be accessible to all. As a result, village landscapes reflected the centrality of religious institutions to colonial New England life. Over time, a multidenominational religious landscape developed, as Anglican, Methodist, Baptist, and other churches joined Congregational and Presbyterian meeting houses in New England towns. Still, churches located at the core of the community claimed preeminence for the worship of God and—in spite of their number and diversity in many town centers—might bespeak a certain unity of focus.[3]

After the American Revolution, central churches remained crucial to many planners' image of the urban landscape. Plans such as L'Enfant's for Washington, D.C., and Woodward's for Detroit provided churches with prominent central sites.[4] Such sites exhibited the public aspect of institutional religions, expressing the place of churches in the community and their significance in addressing public life. A central location seemed appropriate in an era when religion was not insulated from politics, economy, or social experience. In the heart of a village, town, or city, churches directly confronted spaces and buildings dedicated to other aspects of life.

Although many early Americans commonly agreed on an appropriate location for church buildings, they showed less consensus as to religious architectural style. Church style was bound up with a larger issue about the significance of materialism and the perceived relation between appearances on the one hand and piety on the other. There were at least two contradictory ways to express religious preeminence. By their very stature, monumental forms could claim precedence over the surrounding city and, with external embellishments and rich interiors, illustrate a congregation's willingness to dedicate wealth to religion. Conversely, however, plain styles and modest dimensions could also powerfully express a Protestant religiosity that conspicuously refused to adopt material standards. The realm of the spirit might transcend rather than compete with the realm of economic striving. Puritan meeting houses had boldly eschewed ornament and icon. This attitude was epitomized by builders of early New England meeting houses who refused to build "popish" towers. Some communities minimized the distinction between churches and other sorts of buildings. By employing the same material and style used for secular structures, church builders might set religion in the midst of life, suggesting if not a comfort with worldliness then a determination to take it on. Equally, plainness worked well as a symbol when it contrasted with the sumptuousness of other buildings, sacred or profane. Under some circumstances, plain style was a way to stand apart.[5]

By contrast with a number of colonial town plans, the original 1830 plan of Chicago revealed the town's commercial basis. It made no provision whatsoever for church location, but laid out a rectilinear grid plan that deviated along the course of the Chicago River but otherwise remained uncompromisingly geometrical. Land along the river would clearly be lined with commercial structures. Beyond this, a public square offered the only real spatial distinction among the town's fifty-eight original blocks. Nothing in the plan suggested which lots would be taken up for residences, stores, warehouses, workshops, or churches. The surveyor for Chicago's town plan gave little consideration to the spatial expressions of religion (see fig. 3).

Moreover, at first city settlers did little to distinguish buildings of worship from other structures they constructed. In the first months and years of settlement, religious groups met in private residences or occupied other secular spaces. The Rev. Isaac Hallam, for example,

recalled his congregation's use of Montgomery's Auction House for Episcopal services in the 1830s: "The walls were covered with plats of towns that were to be, and we used to go early in the morning and turn them face to the wall, so that the attention of the people might not be directed to worldly business. . . . I preached from the auctioneer's desk, where during the week town-lots were bought and sold for five dollars apiece."[6] In early Chicago, Sunday religion rubbed elbows with weekday speculation and material striving.

Even where they built separate religious structures, early church builders worked within a local, unspecialized form of vernacular architecture. In 1833, the Presbyterian, Methodist, and Baptist congregations joined in the construction of a two-story wood-frame building at the corner of Franklin and South Water streets. The upper story was used as a school while the congregations shared the first story for Sunday services. Religious use, rather than architectural form or location, set this wood-frame structure apart from neighboring domestic and commercial buildings. This was also true of many other church buildings constructed in the 1830s and early 1840s. The wood balloon-frame structures built throughout the town for residences and businesses were also built to serve as houses of worship. In 1842, the Second Presbyterian Church commissioned local builders to erect a 40-foot by 60-foot "balloon building with square timber foundation set on blocks . . . all to be finished off in good plain style."[7] Plainness seemed fitting for frontier religious and secular buildings alike.

Yet settlers soon made clear that they worshipped in plain style churches from expedience rather than conviction. Simple balloon-frame structures would house congregations only until they attracted sufficient members to build more expensively, like the churches in which congregants had worshipped before moving to Chicago. After gathering a stable and prosperous congregation, Chicagoans set out to replace their first churches according to the familiar ecclesiastical forms of the older towns in the eastern United States. In June 1837, for example, after nearly three years of occupying rented quarters in commercial buildings, the St. James congregation dedicated a $15,000 brick church. Although only 40 feet by 60 feet, the building had Gothic-style windows and door, an organ, bell, carpeting, and a mahogany pulpit that alone cost $2,500. Among Chicago's few hundred buildings, St. James represented an "imposing structure" indeed.[8]

For St. James's vestry members and their families, the building's prominence in size and site was reassuringly familiar. John H. Kinzie, for example, had donated two lots for the building out of his extensive north-side tract of landholdings. Although he had spent much of his childhood on the frontier with his fur trader father, Kinzie had encountered the architectural forms of a developed religious landscape during extended visits to Detroit. More pertinent was the experience of Juliette S. Magill, Kinzie's wife and an active congregant at St. James. She had grown up in Middletown, Connecticut, where a Congregational church had stood prominently on the village green; since 1752, Middletown's Episcopalians had worshiped in a frame church, 50 feet by 36 feet, with a towering steeple.[9] Similarly, vestry member Gurdon S. Hubbard was born and raised in Windsor, Vermont, where a neoclassical, steepled Congregational church had overlooked the town common ever since 1798, four years before his birth. Hubbard came from a family with a church building tradition: In 1822, his uncle, a senior warden in the Windsor Episcopal church, had contributed a large part of the funds for building the architecturally distinguished St. Paul's.[10] Finally, Chicago's St. James church received its name from its first rector, Isaac Hallam, whose New London, Connecticut, congregation had worshiped in an ornate 1780s building topped by a cupola and fitted with the distinctive round-headed windows of eighteenth-century Episcopal architecture.[11] Although Chicago's St. James did not replicate any single eastern church, when it opened for worship in 1837 it restored images of religious monumentality and distinction familiar to settlers from the eastern states.

Other congregations delayed substantial building projects until after the economic depression of the late 1830s. Then, in the 1840s, Chicago's leading congregations built distinctive churches on Washington Street. The earliest congregations located on corner lots facing the south side of the public square, the block bounded by La Salle, Randolph, Clark and Washington streets. Washington Street enclosed the side of the square farthest removed from the busy city blocks near the river. Here the First Presbyterian, First Baptist, and First Methodist Episcopal churches formed a religious precinct adjacent to the only distinctly civic block in the original plan of the city. The First Unitarian and First Universalist churches soon located themselves on Washington Street one block to the east of the square; the Second Presbyterian built two blocks east on Washington Street. By the end of the 1840s, Washington Street stood out as Chicago's "street of churches"[12] (figs. 40 and 41).

These six congregations had not coordinated their moves to Washington Street. Yet collectively they created an image of religious primacy clearly derived from an established tradition of American town form. Here along Washington Street city builders juxtaposed religious and civic authority. Churches faced commercial buildings on surrounding blocks while the public square provided a grand and distinguished setting. The landscape link forged between the public square and the churches reflected the organization of many settled towns of the eastern United States. It reflected, too, a frame of mind about religion and its place in community. Sited on open ground, churches oversaw surrounding space and were made visible to pedestrians and riders who traveled the nearby streets; religion was given a realm to command. Clustered together, the churches along Washington Street made a claim about the nature of the city as a whole.

For most congregations, moreover, the move to Washington Street coincided with construction of monumental church buildings. The First Methodist Episcopal Society built the first church on the street, a modest balloon-frame structure located at the southeast corner of Washington and Clark streets, diagonally across the intersection from the public square. Yet within a few years the congregation replaced it with a substantial neoclassical brick church designed by pioneer architect and builder John M. Van Osdel. The basement meeting and Sunday school rooms were enclosed by stone walls. The upper floor's sanctuary auditorium rose thirty feet and was enclosed by brick walls. Brick Doric-order pilasters embellished the corners and the spaces between the windows. A belfry, clock tower, and steeple that rose 105 feet over the roof line gave the church exterior its most notable element. Van Osdel, the son of a Baltimore master carpenter, had closely studied architectural pattern books and builders' guides at New York's Apprentice Library, then moved to Chicago and entered the architectural profession in 1836.[13] First Methodist bore the influence of Van Osdel's pattern book training.[14]

In the course of the 1840s, other congregations that settled along Washington Street either replaced their plain wood-frame structures with brick or else added new embellishment to them. In 1845, the First Unitarian Church added a four-stage Doric order tower and spire to its 1840 wood-frame building, 42 feet by 60 feet, constructed on Washington Street. The building echoed the form of the nearby wood-frame church, 30 feet by 45 feet, topped by a cupola and tower built for the First Universalist in 1844 on the other side of Washington Street. Both buildings had Doric pilasters as well as classical columns placed in antis on either side of the main entrance. In 1844, the First Baptist congregation also completed its church, a brick 50-foot by 80-foot struc-

40. Panoramic view of Washington Street churches and City
Hall-Courthouse on Public Square, 1857. Detail of the J. T.
Palmatary/Braunhold and Sonne panorama of Chicago.
Chicago Historical Society, ICHi 05655.

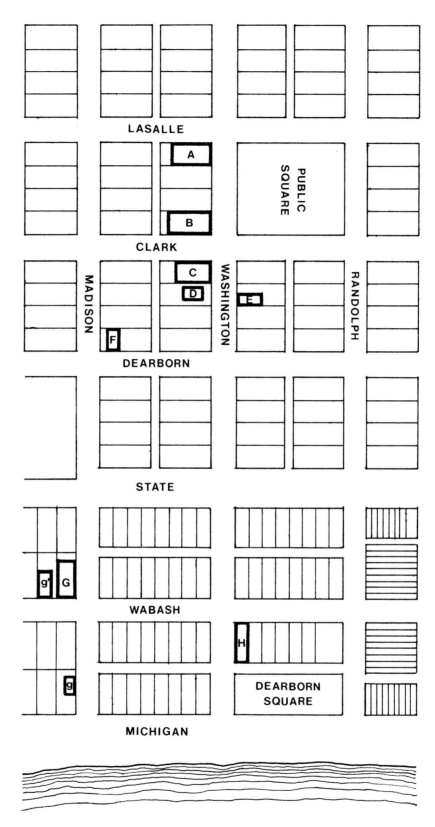

41. Site plan of Washington Street churches, 1857. This map identifies the churches visible in the detail of the J. T. Palmatary/Braunhold and Sonne 1857 panorama of Chicago. The churches are the First Baptist (A); First Presbyterian (B); First Methodist (C); First Universalist (D); First Unitarian (E); Grace Episcopal (F); St. Mary's (G); and the Sisters of Mercy Convent (g'); the Catholic Bishop's Residence (g); and the Second Presbyterian church (H). Map by Areta Pawlynsky. (G, g', g not visible in fig. 40.)

ture, at the southeast corner of Washington and La Salle streets and facing the public square. Here a 112-foot-high tower and spire rose above a six-column Ionic portico. When this structure burned in 1852, the Baptists rebuilt, combining classical, colossal-order pilasters with a 170-foot spire (fig. 42). Next, the First Presbyterian took up the opposite Washington Street corner, at Clark Street, also facing the public square. In 1849, the congregation built a "tasteful structure of the Doric order of architecture" designed by Van Osdel. The brick building, 66 feet by 113 feet, was topped by a tower and steeple that reached 163 feet (fig. 43).[15]

Following these building campaigns of the 1840s, churches stood out as one of the most prominent features of Chicago's landscape. "Chicago may be emphatically pronounced the city of churches," declared a newspaper in 1849. "To a traveler viewing it from a distance it presents the appearance of a congregation of spires."[16] (See fig. 44.) Through the height of their spires, their distinctive settings along Washington Street, and their monumental form and style, churches dominated both the skyline and landscape of the city. The location and designs for Washington Street churches established religion as both central to and distinct from Chicago's commercial and domestic life. Churches shared their neoclassical style with the city's first courthouse and some early residential structures. However, the addition of towers, steeples, and spires set them apart from these other buildings. "The spire has come to be a recognized sign of religion," said one 1850s pattern book for churches in the West. The spire indicated that a church was "not a barn, nor a dwelling-house . . . a merchant's exchange, a bank, a town-hall, or a school-house, but a building of a different character."[17] This information could hardly have surprised aspiring church builders. Nonetheless, it reminded them of the important role of architectural differentiation as a public revelation of religion and a symbol of refined urban culture. Chicago's church-building elite could fairly feel that it had successfully established an appropriate religious landscape, one that followed from the assumptions that shaped established eastern centers and that testified to its own character as a pious and civilized community.

Even so, Chicago's religious landscape rapidly and dramatically shifted. The first token of change came in 1849 when the Second Presbyterian congregation started to build the last of the major downtown churches on Washington Street. Located on the northeast corner of Washington and Wabash, this church shared some common features of the 1840s churches. It also anticipated the rather dramatic reorientation of the religious

42. First Baptist Church, Washington Street, unknown architect, 1852. Photographed from the dome of the City Hall and Courthouse in 1858. Chicago Historical Society, ICHI 05730.

43. First Presbyterian Church, Washington Street, John Van Osdel, architect, 1849. Photographed from the dome of the City Hall and Courthouse in 1858. After the First Presbyterian congregation moved to Wabash Avenue in 1857, the church was used for a time as a Mechanics Institute and then demolished to provide a site for a commercial building. At the left in the background is the U.S. customhouse under construction. Chicago Historical Society, ICHi 05732.

44. Church steeples on Chicago's skyline, ca. 1849, looking forward toward downtown from near corner of Chicago Avenue and State Street, from "View of Chicago as seen at the top of St. Mary's College" by August Hermann Bosse. I. N. Phelps Stokes Collection; Miriam and Ira D. Wallach Division of Art, Prints and Photographs; the New York Public Library; Astor, Lenox and Tilden Foundations.

landscape that would take place in the 1850s and 1860s. As it modified the pattern set by the other Washington Street churches, it introduced strikingly modern elements of religious architecture. Because of its pivotal nature, the Second Presbyterian warrants a more extended view.

The Second Presbyterian congregation was founded in 1842 and first worshipped in a city saloon, then in the Unitarian Church on Washington Street, and finally in its own modest balloon-frame, "plain style" building on Randolph Street near Clark Street. Originally having only twenty-six members, the congregation waited for increased numbers and wealth to begin construction of an "elegant piece of architecture" at the northeast corner of Washington and Wabash.[18]

The site for the new church, overlooking Dearborn Park, was considered "one of the finest in the city: upon the Lake shore . . . having three sides—south, east, and west—fully exposed to view."[19] Thus, like other Washington Street congregations, the Second Presbyterian chose a site that visually took advantage of the expansive setting provided by open public space. Yet the way in which the church was situated in relation to the park is significant. Dearborn Park, formed in 1839 as part of the government's sale of Fort Dearborn, was indeed public ground, but it was far less closely associated with public authority than was the central, public square. In addition, Second Presbyterian did not address its public ground directly: its rear facade, rather than its front, faced the park while the entrance opened onto Wabash Avenue, toward the settled part of the city. Fronting onto the street was a reasonable decision; it would have been awkward to face the park, since beyond it lay only Michigan Avenue and Lake Michigan. Second Presbyterian did not share the propinquity of the other Washington Street churches with the public square.

Equally important, Second Presbyterian departed from the prevailing neoclassicism of the other Washington Street churches. Initially, the building committee had commissioned Van Osdel, who designed an Ionic-order building topped by a spire "superior to anything of its kind in the city."[20] Then the committee had a change of heart; it disapproved the design and turned to New York architect James B. Renwick, who provided Chicago with its earliest authoritative Gothic revival design.

Indeed, Renwick's Second Presbyterian revealed a much closer study of Gothic architecture, both interior and exterior, than had the earlier symmetrical Gothic design of St. James. The church trustees turned aside appeals for economy and insisted on building in limestone rather than brick. The building, 80 feet by 110 feet,

included both a bell and a clock tower on the corners on the front facade, one 161 feet high, the other 64 feet (fig. 45). The nave provided space for 144 pews, separated from side aisles by columns and pointed arches. The church covered ground nearly twice the size of the congregation's original building, and the pews in the nave and the galleries combined provided three times as many seats. Carved wood beams with fretted woodwork supported the roof, "the whole in imitation of the oak roofs of some of the famous old English cathedrals."[21] In the place of the clear-glass windows of the 1842 structure, the new building included stained-glass tracery windows manufactured in New York, separated on the exterior by massive stone buttresses.

It is difficult to detect any fundamental difference in the background, religiosity, or ideals of the founders of this congregation that might explain their abandonment of neoclassicism and of the conventional orientation of Chicago's other leading churches toward Washington Street and the public square. A brief look at Second Presbyterian's founders and officials reveals commonality with others of Chicago's church-building elite. Benjamin W. Raymond, who as Chicago's mayor had worked to establish Dearborn Park in 1839, helped select the church site nearby a decade later. Raymond's father had been a city builder before him. As agent for a group of New York City land proprietors, Raymond senior had founded Potsdam, New York, in 1803; he sold the settlement's land and water privileges, built dams and mills, laid out the roads and streets, and shaped the city's public spaces. Born in 1801, the son watched his father and other settlers carve Potsdam out of the St. Lawrence County wilderness. Raymond senior, led Sabbath services in his own house in the early years. Then in the 1810s and, more especially, in the 1820s, Potsdam joined other upstate New York towns in church building campaigns, the physical embodiment of contemporary religious revivalism. When the First Presbyterian congregation completed a church with a one-hundred-foot spire in 1822, Potsdam had its first church building. Religion mattered to the younger Raymond—living in Rome, New York, in 1826, he participated in a month-long revival crusade led by evangelist Charles G. Finney. After the revival he joined Rome's First Presbyterian congregation, worshipping in a spired building adjacent to the public square.[22]

Many of Second Presbyterian's other founders were also second-generation church builders, men who were firmly grounded in an ideal of a stable and monumental religious center. Trustees John Chandler Williams in

45. Second Presbyterian Church, James Renwick, architect, 1849. This view along Washington Street shows the residential character of the blocks adjacent to the church in the 1850s (photo ca. 1861). Courtesy of the New-York Historical Society, New York City.

Washington County, New York, and Hadley, Massachusetts; William H. Brown in Auburn, New York; and Thomas Butler Carter, son of a carpenter-church builder, in Madison, New Jersey, had grown up in families that built and worshipped in neoclassical churches that, through their high style, steeples, and prominent sites dominated their local communities.[23] Yet for all that they did significantly reconfigure the urban religious landscape. The changes introduced by Second Presbyterian would not seem significant were it not for the fact that so many other Chicago congregations quickly followed its example. In the 1850s and 1860s, many churches abandoned their prominent locations along Washington Street and in the city center. In 1855, only six years after having built grandly on the public square, the trustees of First Presbyterian sold their Washington Street church and abandoned downtown, relocating to a new church completed in 1857, three-quarters of a mile southeast on Wabash Avenue between Van Buren and Congress streets (fig. 46). In 1857, the First Universalist also moved to a new church on the same block. These churches were followed to Wabash Avenue by the First Unitarian in 1863 and the First Baptist, which sold its Washington Street building in the mid-1860s to build on Wabash Avenue. During these decades, moreover, churches from other parts of the city relocated on Wabash. The Wabash Avenue Methodist Episcopal Church built at Wabash and Harrison (1857); Grace Church moved to Wabash and Peck (1857), then to Wabash and 14th (1868); the Plymouth Congregational Church built at Wabash and Eldridge Court (1867) (fig. 47).

While Wabash quickly replaced Washington as the "street of churches," still other congregations left downtown for other sites. Chicago's Catholics had followed much the same pattern as the Protestants. In 1833, they built St. Mary's, a modest balloon-frame structure topped by a small bell cot, at the corner of Lake and State streets. Ten years later, mason Peter Page and carpenter Augustine Taylor built a brick church for St. Mary's at the southwest corner of Wabash and Madison, one block south of Washington Street. The building was designed in the neoclassical style of the Washington Street churches and had a portico supported by four Ionic columns and an Ionic-order tower topped by a spire. Although the Protestant churches were freestanding and independent, St. Mary's had ambitious plans to serve as the center of a cluster of Catholic institutions. "There is a lot of ground adjoining the new church upon which may yet be erected the diocesan Cathedral," wrote Bishop William Quarter in 1844. "There is also a lot in the rear of the

46. First Presbyterian Church, Wabash Avenue, W. W. Boyington, architect, 1856 (1857–71 photo) Chicago Historical
Society, ICHi 22332.

47. Map of the relocations of churches away from Washington
Street and downtown to Wabash Avenue, 1856–67. First Baptist (A, moved 1865), First Presbyterian (B, moved 1857), First
Methodist (C, remains on site), First Universalist (D, moved
1857), First Unitarian (E, moved 1863), Grace Episcopal (F,
moved 1857), Catholic Cathedral (G, relocated from St. Mary's
parish to Holy Name parish in 1859), Second Presbyterian (H,
built 1849), Trinity Episcopal (I, built 1861), Plymouth Congregational (J, built 1867), Wabash Avenue Methodist (K, built
1857). Map by Areta Pawlynsky.

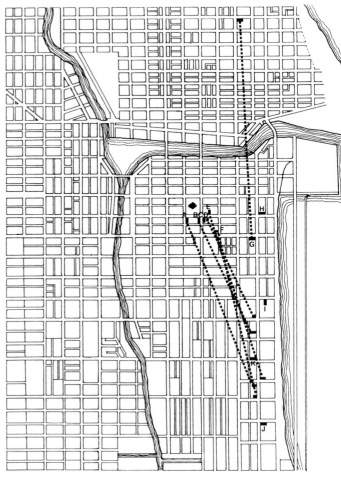

church, where a free school for the poor of the congregation may in course of time be erected."[24] The diocese took the existing St. Mary's as its cathedral rather than add another structure, but within a few years Bishop Quarter and the Sisters of Mercy founded St. Francis Xavier Academy on the site. The sisters also established a convent and school in a four-story brick building, forty feet square, designed by Van Osdel and located on the lot adjoining St. Mary's. They moved the old frame church behind St. Mary's to serve as a free school and added clerical housing as well. By the late 1840s, St. Mary's effectively enjoyed the same landscaped setting as the Protestant churches that fronted on public grounds; the bishop's house occupied a large, landscaped site directly east of the cathedral, so that an image of clerical authority replaced the civic authority of the public square. The complex (fig. 48), which had yet another school building added in 1865, had "a most beautiful prospect of the broad Lake Michigan lying out before it," east of the block occupied by the house.[25]

Soon, however, the diocese began seeking headquarters outside downtown. In 1859, newly appointed Bishop James Duggan designated Holy Name Church, on the north side of the river at the corner of State and

48. St. Mary's Cathedral (at right), Wabash Avenue, Peter Page, and Augustus Taylor, builders, 1843; St. Xavier's College and Sisters Of Mercy Convent (center), John Van Osdel, architect, 1847, addition (left), 1865 (1865–71 photo). Chicago Historical Society.

Superior, as the cathedral. St. Mary's Church, the convent, schools, and bishop's residence remained downtown, but the focus of the diocese shifted away from the
center. Holy Name Church stood adjacent to the University of St. Mary of the Lake, founded in 1844, which was
considered "pleasant, healthy, and sufficiently remote
from the business part of the city to make it favorable to
the pursuits of study."[26] Between 1844 and 1849, the
neighborhood, a mile northeast of the courthouse, developed as an area for residences, sharply contrasting with
the business structures filling the blocks around St.
Mary's. Founded in 1848, Holy Name had occupied a
frame structure. In 1853, the parish started constructing
a monumental Gothic-style brick structure, 84 feet by
190 feet. The bishop left St. Mary's for the largest church
in Chicago.

In the early 1850s, worshippers at St. James Episcopal
Church had added a tower, stained-glass widows, chancel, and galleries to their 1837 structure, located on a
busy block adjacent to the Chicago River. Within the
decade, however, the congregation moved away from
the city center to a residential area just a block south of
Holy Name Cathedral. Designed by Edward Burling, the
new St. James, 72 feet by 148 feet, was twice the size
of the old church and made more sophisticated use of
Gothic forms (fig. 49). Holy Name and St. James were
only two of the numerous churches that left downtown
but did not cluster along Wabash. Their relocations were
part of a larger and dramatic reordering of Chicago's
religious landscape.

What happened in the city center that shifted the
traditional perception of that place as the appropriate
site for religion? What made the outlying sites or different centers of population, such as Wabash Avenue,
seem more appropriate? The relatively new notion of
distinct *centers* within the city for residence and business represented a novel conception of urban form quite
different from the view that prevailed in the 1840s. How
did the changing structure of both urban space and religious congregations redefine the nineteenth-century
religious landscape? Understanding what lay behind this
abandonment of the downtown area requires looking
first to commercial building, then to the development of
residential realms within the city, and finally to emerging
connections between churches and domesticity.

In the simplest sense, the churches moved because
of downtown commercial development. Between 1850
and 1870, Chicago's population increased from 29,963 to
298,977. The accompanying expansion of commerce
transformed the city's landscape. At the center of the city,

49. St. James Epsicopal Church, Edward Burling, architect,
1856 (photo ca. 1865). Chicago Historical Society, ICHI 22327.

the more intensive commercial use of land and larger scale commercial buildings accompanied Chicago's rise from county seat to regional metropolis. During the 1850s, individual commercial buildings surpassed religious structures in both scale and cost. In 1857, for example, as St. James, Holy Name, and the First Presbyterian were being built on budgets of $90,000 to $100,000, buildings such as the Richmond House Hotel at Michigan Avenue and South Water Street cost more than $120,000. Commercial interests increasingly built with brick and masonry, materials previously reserved primarily for churches. During the 1850s, stone and cast-iron fronts gave commercial buildings additional architectural distinction. John M. Van Osdel's 1856 design for Allen Robbins's $95,000 South Water Street building, with its two Daniel D. Badger and Company cast-iron facades, was viewed as a Chicago monument, "one of the largest and finest iron front buildings in the country, and an ornament to the city."[27] During 1856, Van Osdel had installed more than three hundred feet of Badger's cast-iron building fronts along Lake Street. These five- to six-story buildings rose over seventy feet from the street. Chicago's churches dominated the landscape a good deal less after the commercial expansion of the 1850s than they had a decade earlier, when three- and four-story commercial structures prevailed.

The rising scale and pretension of commercial buildings led some Chicagoans to feel that commerce was encroaching on their churches. In 1868, George S. Phillip in *Chicago and Her Churches* observed that the "centres of population and business in the city were undergoing such important changes" that "the place of worship must also be changed."[28] In 1865, Zephaniah Humphrey, minister of First Presbyterian, similarly explained why his congregation had moved away from Washington Street a decade earlier: "It was found that the location was not good, the surrounding population being driven away by the encroachment of business, and the place becoming constantly more dusty and noisy."[29] The church's new site, three-quarters of a mile southeast on Wabash between Congress and Van Buren streets, was "a much more retired place of worship."[30] In a dramatic transfer of religious property to commercial interests, the First Baptist Church sold its Washington Street building to the Chamber of Commerce in the mid-1860s. An 1889 church history insisted that "the advancing tide of commercial enterprise, levying new demands upon the central portions of the city . . . was heard surging against the walls of the old edifice, and admonishing the church that the spot, so long hallowed by innumerable tender and sacred associations must soon yield to the march of

events."[31] Ironically, even Second Presbyterian, with its monumental Gothic structure and Wabash Avenue site, began looking in 1866 for a new location "on account of the rapidity with which the church was being surrounded by business structures."[32] (See fig. 50.) Indeed, in 1868 the Lord and Smith Company, which was building a five-story masonry store and warehouse on Wabash Avenue, sought permission to build on an eight-to-ten-inch strip of church property.[33] The mass of Lord and Smith's building and the adjacent commercial structures on the block of Wabash Avenue dwarfed the church building (fig. 51; see figs. 45 and 50). In response, the church resolved to move south to the corner of Wabash Avenue and 20th Street.[34] In the growing downtown area, then, many churches seemed dwarfed, while new commercial structures filled vacant lots that had created a prospect surrounding them. Church planners abandoned one part of their inherited tradition—central location and association with public authority—to retain another—monumental form and visual domination achieved through a size and style that depended greatly upon the form of adjacent structures.

As they felt pushed by commerce, churches also felt pulled by the creation of new residential areas in the city. Settling along Wabash in the 1850s and 1860s meant moving into the primary elite residential neighborhood that had developed in the wake of commercial expansion in the downtown area. When the enumerators of the 1860 federal census visited Wabash and Michigan avenues between Madison Street and Park Place, they found most of Chicago's wealthy residents. Here native-born men working as lawyers, judges, bankers, real estate brokers, commission merchants, and manufacturers headed the households, employed numerous domestic servants, and often owned hundreds of thousands of dollars of real estate and tens of thousands of dollars of personal property (fig. 52).[35]

The line of eleven three-bay wide row houses making up "Michigan Terrace" were the most notable residences of the 1850s along Michigan and Wabash avenues. The first three Wabash Avenue churches built during the 1850s, the First Universalist, First Presbyterian, and Wabash Methodist Episcopal, were located on three blocks of Wabash Avenue adjacent to this development (fig. 53). Michigan Terrace exemplified the rise of class-segregated neighborhoods. In the 1850s, the Michigan Terrace project brought together several wealthy Chicago residents, including lawyer Jonathan Young Scammon, real estate broker Tuthill King, railroad president Charles Walker, hotel proprietor Francis C. Sherman,

50. Second Presbyterian Church, James Renwick, architect, 1849, and Crosby's Opera House, W. W. Boyington, architect, 1865 (view ca. 1868). Crosby's stood on the north side of Washington Street between Dearborn and State streets and contributed to the commercialization of Washington Street, deemed problematic by many congregations. Chicago Historical Society, ICHI 22306.

51. Wabash Avenue, north from Washington Street, Second Presbyterian Church (at right), James Renwick, architect, 1849 (1868–71 photo). These commercial structures intruded on the setting of the Second Presbyterian Church. Chicago Historical Society, ICHI 22315. (See fig. 45.)

52. Wabash Avenue, looking south from Eldridge Court (photo ca. 1870). Chicago Historical Society, ICHi 22336.

53. Michigan Terrace, W. W. Boyington, architect, 1856 (photo ca. 1870). Looking west from the lakefront, the tower of the First Presbyterian Church, Wabash Avenue, appears behind Michigan Terrace. Courtesy of the New-York Historical Society, New York City.

newspaper publisher William Bross, lumber dealer John Sears, Jr., and, slightly later, commission merchant P. F. W. Peck. These men owned all of the lots facing Michigan Avenue between Van Buren and Congress streets. Despite the prevailing pattern of detached houses among Chicago's elite, they planned a unified brick terrace. Their joint action would ensure architectural harmony and also preclude other buildings or residents from settling unexpectedly in their midst.

It took some time for these land owners to reconcile their visions of the terrace project, and "meanwhile, the owners becoming more wealthy, and the location more desirable, it was conceded by all the parties interested that nothing less than palatial marble fronts would comport with the value of the site and the style of building prevalent in th[is part of the] city." William W. Boyington designed the symmetrical terrace with fourteen four-story row houses; the two end houses and the two middle houses had an additional floor. The 1857, financial panic interrupted building before the final three houses were constructed. Nevertheless, one review declared it "one of the most beautiful blocks of private dwellings which any city in the Union can boast. Its graceful proportions and harmonious style of architecture indicate the good taste and superior ability of the architect, as the elegant workmanship does the unsurpassed skill of Chicago mechanics."[36] Residents could hope that Michigan Terrace would also speak to their personal taste and refinement.

In Chicago the changing spatial structure of commerce and residence combined with the prevailing voluntary and uncoordinated basis of religious worship to foster the removal of leading churches from the center of the city. Protestant denominations did not share any consistent strategy for when and where to build churches. Unlike the established churches of colonial America, these congregations were voluntary organizations. They relied on the pew rent and contributions of members and contributions to support all religious activities and building campaigns of the congregation. They competed for wealthy members who could support a talented minister, professional choir, or sumptuous building program. Unlike the Catholic church, which maintained geographically defined parishes and shared resources among churches, Protestant congregations had no clear geographical limits. Officials of the Washington Street churches faced the possibility that outlying churches, with new wealthy members, could threaten their denominational leadership. When, for example, Trinity Church moved farther from downtown, the rector wrote that its neighborhood was "gradually being absorbed

into the great vortex of commerce, and private residences were rapidly moving southward and westerly, and fashionable churches were offering the further attraction of greater proximity for one and another." By moving, the parish hoped to be "less exposed to the injurious influences of a fluctuating, shifting social condition": and to "secure a greater measure of that permanence and stability that are so essential to parochial prosperity."[37] By leaving Washington Street and following their wealthy members to the emerging residential enclaves, many churches sought continued denominational leadership. In other words, they followed the monied.

The extremely short period of organized church worship encouraged the transience of Chicago's congregations. In abandoning their central sites, Chicago church members revealed little of the sentiment or solicitousness for established tradition that had earlier led them to build by the public square and along Washington Street. In retrospect, the First Baptist's historian romantically described its Washington site as "hallowed by innumerable tender and sacred associations," but in fact the congregation had worshipped on the site for less than two decades. Churches that had been occupied for ten or fifteen years obviously lacked the rooted significance of buildings used for family rituals over generations. And if there was little reason to stay, there was often ample impetus to move. Chicago's growth had substantially raised land values, and evidence abounded concerning the financial rewards of unsentimentality in relation to land. The value of Chicago land rose sixtyfold between 1830 and 1836, from $168,800 to $10,500,000, and in the latter year, at the height of city speculation, some Chicago lots exchanged hands several times a day. In the late 1860s, real estate interests offered the Second Presbyterian $195,000 for the land and building that had cost only $42,000 twenty years before. Sharp rises in land values and a pervasive view of land as a marketable commodity prompted many church relocations.

For all that, it is important not to overstate the purely economic advantages of church relocation, discounting the texture of what was a complex phenomenon. Congregational moves were undoubtedly mystified: When church historians attributed relocations to the surging tide of business enterprise, they elided economic calculus and human choice to invoke an impersonal course of events. Some of the concrete reasons they offered for relocation, moreover, sound insufficient; if dust and noise afflicted downtown on weekdays, for example, these inconveniences abated on Sundays, when most congregants came to worship. Similarly, rather small distances separated residential enclaves from the business center in the 1850s, and churches in the former were only slightly more convenient than churches downtown.

Yet if official explanations of church relocation were often superficial, they were not merely cloaks for self-interest. Money netted from the sale of downtown lots, after all, did not line the pockets of church officials who decided to move; the "profits" of congregations underwrote the construction of new monumental, ornate houses of worship. The new churches could provide congregants with status and boost the value of their newly developed residential neighborhoods. In the 1860s, just as expensive church buildings were rising along Wabash Avenue, Frederick Law Olmsted observed that "scarcely anything is done publicly in Chicago entirely free from the current of business interests." He confided to his journal: "A church steeple will be built higher that it otherwise would be because the neighboring lot-holders regard it as an advertisement of their property. There is nothing done in Chicago that is not regarded as an advertisement of Chicago."[38] Nevertheless, churches worked as advertisements only because of the powerful cultural notions that assigned religious symbolism an important role in fostering a cultivated, refined, and prosperous neighborhood. It is precisely those notions—the belief that churches properly enhanced residential areas of a city—that need exploring.

There was nothing inevitable about the departure of churches from their central locations in Chicago and other American cities. Given the impassioned calls for religion to keep pace with commerce, churches could have usefully challenged the commercial preemption of the most visible, accessible, and valuable land in the city. Yet church relocation was part of the formation of an urban domestic realm to which some park plans also contributed. Like parks, churches located outside the business district underscored a separation between commerce and the rest of life. They contributed to the emergence of a domestic arena devoted to refined sensibilities and lofty ideals, to piety and nurturance, to family ties and, in fact, to domesticity (fig. 54).

The creation of new residential enclaves accompanied new, culturally imposed divisions between men and women, articulated most clearly in an ideal of womanly nature and activity. The precept that ladies were to be responsible for the home and gentlemen for the marketplace was central to well-to-do Americans' identity as a class. Historians have emphasized that the ideal of *separate spheres* represented a class ideology, fraught with

contradictions. It largely ignored the lives of working-class women, and it did not even represent accurately the actual experience of middle-class females—the women who mattered to bourgeois men.[39] Yet the "cult of domesticity" remained a powerful ideology that deeply affected nineteenth-century Americans and their landscape. Among other things, it made it seem appropriate for churches to associate with residence rather than a more public sphere (fig. 55). Weaving an increasingly close association between religion and the home, nineteenth-century city and culture builders redefined the meaning of both institutions.[40]

In this respect, church relocation represented the feminization of urban religion. Women outnumbered men in the membership of Chicago's transient churches. Arguably, the concern of church officials over the dust, noise, and travel to downtown (however short) might have expressed deference to ladies' supposed sensibilities and frailties. It was hardly "agreeable," said one writer, to "wind one's way to the sanctuary through barricades of petroleum, molasses, salt, tea, and fish barrels, empty soap boxes, stacks of heavy hardware and such worthy but worldly minded institutions" as lined the downtown streets.[41] If men found intimations of commerce to be unsettling, then surely the finer and frailer sex might also react to downtown that way. Moreover, the call for men and women to inhabit separate spheres involved an emerging sense of propriety that limited women's presence and mobility within the city. When women worshipped in churches located in residential areas, they stayed close to home.[42]

Equally significant, when churches stayed close to home, it may have been easier to confine religious precepts to the domestic realm, easier to debar religion from the marketplace. As historians have noted, the nineteenth-century's idealization of domesticity might have served to rationalize men's pursuit of Mammon in the marketplace by confining morality to a domestic realm that was the responsibility of women, social beings with limited power to effect changes in the male world of work and politics. Women's work (the work of being leisured, decorative, nurturing, pious, and moral) might spare men the burdens of self-doubt, morality, and responsibility for culture while they struggled in the marketplace.[43]

Seen in this light, church relocations surely represented a retreat before Mammon, a yielding of the center that might work simply to remove embarrassing symbols of morality and piety from the consciousness of men at work. This attitude toward churches was apparent in the

54. First Universalist Society's St. Paul's Church, Wabash Avenue, W. W. Boyington, architect, 1856 (1857–71 photo). Chicago Historical Society, ICHi 22313.

55. View southeast from dome of City Hall and Courthouse, late 1860s (photo ca. 1868). In the foreground, the building with the advertisement for Steinway pianos stands on the Washington Street site of the First Presbyterian Church (1844–57); across the street to the left is the First Methodist Church block (built 1857). In the distance, three large churches are visible; at the left the double cupola of Trinity Church; to the right, on Wabash Avenue, is the single steeple of St. Paul's; and, to the far right, is the turreted tower of the First Presbyterian Church. Chicago Historical Society, ICHi 05744.

writings of city boosters who hailed the religious land-scape only as it supported material ends. If, as some apologists' "law of development" stated, fine houses of worship and related religious institutions rose on the foundations of material well-being, then why not point to fine churches as evidence of prosperity, useful in attracting more investment and promoting further growth?[44] Churches (like parks) provided "a capital gauge for the measurement of the progress and prosperity of the city."[45] Thus, in 1873, proudly presenting lithographs of Chicago's finest churches, a real estate and building journal called the *Land Owner* reported:

> If it be true that the church architecture of a city gives a clue to its commerce and trade, then certainly our readers abroad can form an idea of what Chicago really is from a different standpoint than has ever been given. . . . Here in this ever bustling, ever moving, noisy, restless city of commerce and greed, it is a relief to pause in the pursuit of gain, and, turning aside from the palaces of trade on every hand contemplate this long array of quiet, beautiful temples reared to the worship of Him who, in His wonderful mercy, has so signally prospered our city.[46]

Remarkably, this account speaks of churches as if they were mixed among downtown commercial buildings, finessing the fact that they had yielded that terrain. In its depiction of churches as advertisements for the city, however, the *Land Owner* described a religion that was simply sentimental, acknowledging and even celebrating religion's inability to encumber commerce. Indeed, the *Land Owner* assigned only a very small part of life to religion—churches occasioned merely a momentary pause in Mammon's day. In this formulation, piety did not compete with marketplace activity, nor did religion function in any way as a critic of ambition or greed. The contemplation of churches offered a restorative moment of leisure; it was like nothing so much as the contemplation of nature in nearby parks. Religion was not counterpoised to acquisitiveness; it was an ally, not a critic, of the acquisitive soul. Here was a religion that had yielded the center in both a literal and a figurative sense. Like middle- and upper-class womanhood, religious landscape appeared domesticated, finer but frail.

Some Chicagoans opposed the flight of the churches from downtown on precisely these grounds. Yet far from all of the advocates of church relocation sanguinely contemplated the marginalization of religion. A variety of writers in Chicago's press voiced concern about the lack of steadfastness in the city's religious impulse and, in-deed, argued for relocation on that account. In 1866, the *Tribune* strongly objected to the vicissitudes affecting houses of worship located downtown. Complaining of a lack of uniformity in Chicago's development that left wood structures next to brick, low buildings next to high, hovels next to palatial residences, the *Tribune* lamented,

> It is a pity that we have not some kind of order or rule for the association of buildings. . . . Here there seems to be no law against any kind of this 'miscegenation,' even though it extends to the making of a billiard hall and a church as next door neighbors. . . . Can we not have some 'rule of thumb' whereby the proprieties of the situation may be respected, and the finer feelings of our natures be saved from such gross violations?[47]

The *Tribune*'s overwrought language betrayed a peculiar depth of anxiety. In view of the recent emancipation of African Americans in the southern states, ongoing contention over the social and political rights of freed-people, and the elaboration of racist ideology, it took a marked predilection for the inappropriate to bemoan the "miscegenation" represented by different building types located side-by-side along a city street. Contrasting church and billiard hall, the *Tribune* revealed a chain of associations, counterpoising feminine and masculine, elite and popular, anticommercial and commercial. Churches and billiard halls competed, if not for men's souls, then for their leisure time, and they engendered markedly different frames of mind. Quite clearly, the *Tribune* expressed the middle- and upper-class desire for a carefully segmented physical and social world, a fragmentation of life and consciousness. Equally important, it voiced an underlying sense that religion might easily be overpowered when juxtaposed with the forces of commerce. In 1858, for example, one Episcopal rector who had served a parish temporarily quartered in a bustling public hall complained of distraction and competition, with "billiard rooms and saloons below us and a concert hall on the same floor, even during our daily services in Passion week we were compelled to listen to the discordant sounds of Negro Minstrels."[48] Downtown, said another commentator, a church might become surrounded and "submerged by the tide of business."[49] To some, relocated churches were preferable because downtown religion seemed on the defensive.

In their desire to be shielded from commerce, some Chicagoans sought a substantial realm that might be set against the world of Mammon downtown. Locating churches alongside residences might work to enlist women's supposed superior morality on behalf of religious institutions. At the same time, churches augmented the authority of the domestic realm over and against the public sphere of material striving. To the extent that men, in the positions of clergy, trustees, deacons, vestrymen, and members wielded power in the church they, too, contributed to this cultural campaign.

What was at stake in all this, in the *Tribune*'s words, was a sense that contact with the commercial and the vulgar constituted a "gross violation" of refined sensibilities. Associations of Mammon, evoked simply by views of commercial buildings, made it difficult to indulge in finer feelings and tender emotions. Like park construction, church construction was intimately bound up with a preoccupation with a cultivated, moral self being created and nurtured in the home, the garden, and the park (fig. 56). If anything, the respite provided by this realm was more essential to middle-class men, who worked downtown, than to middle-class women, who generally did not.[50] To Chicago's city builders, in other words, certain centers of the city increasingly seemed appropriate to certain identities and consciousnesses, and respectable women's presence at church, as along the promenade, marked it as a particular social setting appropriate for *men* to indulge and develop their "finer" sensibilities. Church relocation, in this view, was not merely a matter of real estate values; it was a matter of securing the individual natures of middle- and upper-class men. The downtown part of the city seemed increasingly inimical to certain valued experiences. If this were so, then to put religion on the periphery might be the only way to make it central.

The reconfiguration of the urban religious landscape suggests the need to modify further an understanding of the nineteenth-century ideal of separate spheres. The men who served as church officials did not limit themselves to business and political affairs. Although they moved church buildings to the domestic realm, they did not fully abandon responsibility for morality and culture to women (any more than they turned over final control of the churches to them). Ultimately, the rationale for separate spheres required that domestic values make themselves felt on men, and some Chicagoans at least seem to have taken seriously the danger that religion, like women, might be trivialized even as it was exalted. Elite concerns over the capitulation to Mammon were real, and some Chicagoans' anxiety over the eclipse of downtown religious buildings by commercial structures stemmed from this awareness. One response was to embrace a contracted awareness, achieved through the removal of churches from the scene of unpleasant associations; another response was the contracted public responsibility in some ways encouraged by voluntary denominationalism. Churches were associated with the influence of the home and its realm of private sensibilities. Yet private sensibilities genuinely concerned Chicago's city builders. It was with some difficulty and tentativeness that they set about fragmenting the consciousness and cityscape alike.

In this framework of cultural meaning and change, the rationale for the adoption by Chicagoans of a new style in church building becomes clear. Architecturally, the new churches of the 1850s and 1860s gave up the neoclassicism of the 1840s for the medieval forms of the Gothic revival. Here again the Second Presbyterian anticipated other city congregations. Just as they followed Second Presbyterian to Wabash Avenue, so other leading churches turned to Gothic forms. Relocation and adoption of a new style were clearly linked: for many leading congregations, the sale of valuable downtown sites provided the money for Gothic revival construction. Built of brick and, more often, stone, Gothic churches eclipsed the wood and brick neoclassical style structures of the 1840s. Beyond expressing religious commitments, Chicago church architecture fueled denominational and congregational competition. A monumental church exterior and sumptuous interior could attract members as well as an eloquent preacher or a professional choir. The profits of downtown land sales funded monumental Gothic church designs competitively aimed at winning souls for the congregation.

The transition from classical to Gothic in Chicago church design represented a provincial manifestation of the broad Anglo-American movement promoting Gothic as the correct form for Christian architecture. In the 1830s and 1840s, English architect Augustus Welby Pugin and English ecclesiologists who worked in the Oxford Movement and the Cambridge Camden Society sought to overthrow the hegemony of classicism in religious architecture. They adopted this goal in the face of what they viewed as a pervasive and contemptible secularism and Enlightenment rationality. By restoring medieval forms, the ecclesiologists hoped to return both spiritual emotion and formal grandeur to Anglican worship. Pugin went further and insisted on a return to pre-Reformation Catholicism.[51]

American Protestants largely overlooked the elaborate liturgical reforms of English ecclesiology. Nevertheless, their Gothic religious architecture was not a superficial, stylistic borrowing. The new style found welcome among those who wanted to demarcate clearly

56. Michigan Avenue looking south from Jackson Street (photo
ca. 1861–71). Chicago Historical Society, ICHi 22334.

city spaces and buildings from one another. Adopting Gothic forms for an entire church building strengthened the distinction between religious and secular forms, and hence between spiritual matters and secular affairs. "The style of every building should be so characteristic that a single glance may be able to decide the purpose to which it is devoted," asserted Frederick Clarke Withers, an important proponent of the Gothic revival in America in 1873.[52] Such calls for clarity of architectural expression challenged the pervasiveness of neoclassicism, in particular Greek revival design. Many architects had embraced the Greek revival as the *American* style in the early nineteenth century and applied it to civic, religious, domestic, and commercial buildings alike. Proponents of the Gothic revival style insisted that medieval forms more appropriately distinguished modern religion from secular commitments than neoclassical forms did.[53] Where some earlier Americans had assumed a unity of culture, a compatibility among republican religion, household, and public life, mid-nineteenth-century city builders saw differences that called out for signification beyond that provided by a spire or steeple.

The enthusiasts of the Gothic revival in both England and America framed their arguments in a highly charged atmosphere of moral purpose and serious misgiving about modern materialism.[54] The Gothic revival evoked a romantic longing for the supposed piety, stability, and community of the Middle Ages that many thought was destroyed by the expansion of commercial and industrial capitalism.[55] This moral tone as well as the formal association of Gothic style with picturesque, organic, and natural forms proved especially attractive to those who viewed nature in transcendental terms as the basis for spiritual regeneration and American nationalism.[56] These ideological and philosophical views of commerce and culture fostered American Gothic revival religious architecture.

After 1850, Chicago's Gothic revival church builders participated in a national enthusiasm for the style. Nevertheless, Gothic revival in Chicago coincided with and drew upon significant local concerns. Most importantly, it coincided with the beginnings of congregational transience. The Gothic style provided comforting images of permanence, stability, and rooted religious tradition in the face of a rapidly changing urban landscape. The Greek revival, of course, had drawn its forms from sources more ancient than the Middle Ages. Yet American architects used neoclassical forms in distinctly modern ways; posing as the heirs of Western civilization, they unabashedly adapted classical architecture for their own

purposes. Americans looking at neoclassical buildings were not encouraged to engage in the historicist fantasy that the structure had stood since antiquity. Gothic style was different. While neoclassicism pointed emphatically forward, the Gothic directed attention backward.

Contemporary descriptions of the neoclassical churches on Washington Street were devoid of references to their antiquity. Starting with Renwick's Second Presbyterian Church design, however, self-conscious formulations of a rooted, medieval past abounded; the Second Presbyterian received the popular name of the "Spotted Church" because its limestone exterior was pitted with pockets of bitumen, giving the structure contrasting light and dark patches (see fig. 45). The masonry strongly reinforced a leading motive of Gothic revivalism, giving the building "a remarkable ancient and unique appearance."[57] Renwick adopted the same limestone material for the Second Presbyterians' 1872 structure, producing "an antique air that cannot but be appropriate."[58] Reviewers of the design for Grace Episcopal Church (figs. 57, 58) on Wabash Avenue assured the public that William Le Baron Jenney designed the building with the "greatest purity" of Gothic style "accurately preserved in all its details."[59] An architecture evoking antiquity and permanence could perhaps compensate for congregational transience. The church designs bespoke a religious tradition predating all the evidence of Chicago commerce and creating a comforting if illusory image of stability.

William W. Boyington, the architect of Michigan Terrace and at least twenty other residences along Michigan and Wabash avenues during the 1850s, also designed the most prominent churches on Wabash. Born in 1818, Boyington had trained in Massachusetts as a joiner, carpenter, and millwork manufacturer, then moved to Chicago in 1853 and became a highly successful architect. In 1856, he designed St. Paul's Church for the First Universalist Society. The Gothic-style rock-faced building had lancet-headed windows and doors and a central tower and wood spire reaching 175 feet above the street, flanked by turrets at the front corners (see fig. 54). The interior was embellished with ribs, purlins, pendants, corbels, and brackets executed in the Gothic style. Following his designs for St. Paul's, Boyington designed First Presbyterian Church (1856); Wabash Avenue Methodist Episcopal Church (1857) (fig. 59); and the First Baptist Church (1866) (see fig. 39). These and other notable churches rising in the late 1850s and 1860s generally cost between $100,000 and $200,000, accommodated between a thousand and two thousand worshippers, and had towers and spires reaching 150 to 250 feet into the air.[60] Theodore V. Wadskier, an architect trained in Copenhagen, also actively participated in the early development of Chicago's Gothic revival. Wadskier designed several major Norman and Gothic style churches after he settled in Chicago in the 1850s, including the First Unitarian's Society's Church of the Messiah (1866) on Wabash Avenue and the building for the Trinity Episcopal Church (1861) located on Jackson Street between Wabash and Michigan. These churches served as models for the churches built in residential neighborhoods, adopting the form of the small Medieval English parish church rather than the form of the great city cathedral.

If a single glance identified these churches and conjured thoughts of the Middle Ages, longer scrutiny revealed more modern accommodations. Architects departed from the "pure Gothic" and models provided by the past in numerous ways. Thus, for example, the English parish church generally stood alone in its churchyard and was visible from all sides. Chicago's Gothic revival churches generally stood on standard lots within the city's larger grid system, with only one side or, in the case of a corner lot, two sides visible. On the city's wide blocks, churches could not easily have parallel street fronts. Moreover, land prices led most congregations to build over their entire lot, leaving little room for landscape embellishment. Constricted sites led Boyington in the First Presbyterian design and Wadskier in the Trinity Episcopal to omit side windows and introduce light to the auditoriums through skylights.[61] The constraints of most building lots forced churches to sacrifice the east-facing altars and south-facing porches that graced their ecclesiastical models. Many congregations also sought to lower the cost of building by reserving stone for the main facades and using brick for side and rear walls. Building budgets sacrificed the "true principles" of Gothic design in spite of prevailing interest in the alleged honesty and integrity of Gothic construction.

Other practical considerations also influenced Chicago's religious buildings. Limestone exteriors with Gothic arches paid quiet homage to traditional public religiosity but masked strikingly modern church interiors. Many congregations required separate rooms and spaces for less formal religious activities such as Sunday school and congregational meetings, social and sewing circles. Many churches also included space for a library, a pastor's study, and even kitchen and toilet facilities.[62] These spaces could be provided inexpensively by locating them in the same building that housed the auditorium or sanctuary. The common pattern placed ancillary spaces on the first floor and raised the auditorium to the second floor, connected to the main entrance by vestibules and a grand staircase. The First Presbyterian (see fig. 43) and First Baptist (see fig. 42) churches on Washington Street had this plan in the 1840s, and Boyington and other architects adapted it to many of their later Gothic revival designs. The plan promoted taller, more impressive exterior walls and hence more substantial buildings, nicely combining deference to modern church use and the desire for a monumental structure.

In church interiors, Chicagoans displayed far greater concern for enhancing their own religious experience and comfort than for remaining loyal to a particular architectural style. American Protestantism drew strongly on the Reformation's emphasis on hearing and seeing the minister, an emphasis that led Chicago congregations to adopt a variety of novel semicircular and graded seating arrangements quite different from seating in the long nave of a parish church. In 1873, reviewing efforts to improve acoustics and lines of sight, one commentator wrote that many Chicago churches had "abandoned the old long and narrow form for the broad semi-circular audience room. The audience is gathered about the speaker, as a crowd gathers itself in open air; not in the form of a brick, with the narrow ends toward the speaker, but in the form of a group that gathers about the old fireplace in mid-winter."[63] Centrally placed pulpits replaced the more reserved settings at one side. In Boyington's semicircular plan for the First Baptist Church, each seat enjoyed "a direct facing to the minister without turning sideways," and the galleries were "low, and graded so that the seats [were] all desirable, and but very little, if any, less so than those on the main floor."[64]

Boyington's graded galleries anticipated the unprecedented introduction of the *dish* or *amphitheatrical* church auditorium. In 1869, many architects and critics laughed at Henry L. Gay's First Congregational Church design (fig. 60)—"the first in the country built on scientific study of scientific principles." Paying close attention to "lines of sight," Gay arranged the pews based on circle segments, "being graduated in such a manner as to diminish as much as possible the distance between preacher and listener."[65] The galleries at the First Congregational Church encircled the entire auditorium (fig. 61). A short distance from the First Congregational stood the Union Park Congregational Church (fig. 62), designed in 1869 by architect Gurdon P. Randall with semicircular rows of pews that sloped gradually downward from the rear toward the pulpit (fig. 63).[66] During the 1870s, the amphitheatrical style enjoyed continued popularity and was used in designs for the First Baptist, First

57. Grace Episcopal Church, Wabash Avenue, Loring and
Jenney, architects, 1867 (photo ca. 1870). Chicago Historical
Society, ICHi 22329.

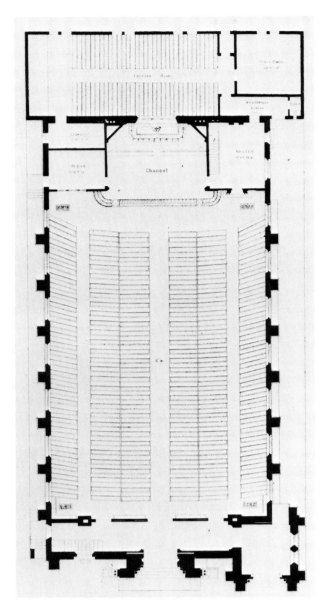

58. Grace Episcopal Church plan, Wabash Avenue, Loring and Jenney, architects, 1867 (1868 plan). Avery Library, Columbia University.

59. Wabash Avenue Methodist Episcopal Church, Wabash Avenue, W. W. Boyington, architect, 1857 (1858–71 photo). Chicago Historical Society, ICHi 22328.

60. First Congregational Church, Henry L. Gay, architect, 1869
(photo ca. 1872–77). Chicago Historical Society, ICHi 22321.

61. First Congregational Church interior, Henry L. Gay, architect, 1869 (photo ca. 1872–77). Chicago Historical Society, ICHi 22323.

62. Union Park Congregational Church, Gurdon P. Randall, architect, 1869 (1906 photo). Chicago Historical Society, ICHi 22303.

63. Union Park Congregational Church interior, Gurdon P. Randall, architect, 1869 (photo ca. 1872–77). Chicago Historical Society, ICHi 22322.

Congregational, First Universalist, Unity, Jefferson Park, Third Presbyterian, Fourth Presbyterian, and Plymouth congregations.

To promote piety, then, church designers mixed medieval religious forms with modern forms of popular culture. Despite the history of tension between religious worship and theater entertainment, Chicago architects working in the Gothic revival style drew upon both religious and secular models, leaving congregations with a "handsome combination of the theatre and the church."[67] The combination of religious and secular that suggested impropriety and danger in a downtown location proved quite acceptable within the confines of the church itself. The installation of the most modern facilities for heat, light, and ventilation extended evidence of the modern world and contemporary form further. Wealthier congregations also enjoyed rich and comfortable furnishings—carpets covered the floors and the pews were cushioned and upholstered. Dark pews of black walnut, black oak, or mahogany frequently echoed ornately embellished open-timber ceilings. In keeping with modern notions of decorative beauty, church auditoriums were elaborately adorned with frescoes of both geometrical and foliated patterns. Otto Jevne, an immigrant Norwegian artist, executed some of Chicago's finest church frescos and stained glass for such churches as Grace Episcopal, First Presbyterian, Trinity Episcopal, and Centenary Methodist Episcopal. The interiors included rich polychromatic paint work, sky blue ceilings dotted with gold stars, and combinations of architectural ornament painted in rich hues of cobalt, ultramarine, carmine, gold, and black. Lavish modern interiors owed a great deal more to prevailing decorative styles than to Gothic precedents.

Although there was tension between Gothic exteriors and modern interiors, there was also a link between them: both were intended to promote religious sensibility, creating a receptive frame of mind in churchgoers and promoting close communion between preacher and worshipers. Chicago's fashionable churches housed a gregarious social life and, often, a liberal Protestant theology. Several elements were involved here. Like parks, the churches of many middle-class Chicagoans invoked nature and solitary reflection on the one hand, society and urbanity on the other. As parks combined references to nature and society, so churches combined the natural and ancient with the artificial and modern.[68]

In interior form, Chicago's Catholic churches generally departed from the Protestant model. Less interested in creating vital contact between worshippers and priest,

Catholics built churches without galleries and left worshippers seated along naves that stretched a good distance from the altar. Holy Name, for example, built in 1853 in a Gothic revival style, measured 84 feet by 190 feet. By contrast, Second Presbyterian, built contemporaneously and also on a grand scale, had an auditorium that measured only 64 feet by 82 feet. Like Holy Name (fig. 64), other important Catholic churches, including St. Michael's and Holy Family, had naves that measured nearly 200 feet long. Different patterns of worship meant that Catholic churches often had a grander scale than their Protestant counterparts.

Moreover, in size Catholic religious precincts generally outstripped Protestant church sites. A commitment to providing schools, academies, convents, parish houses, and related buildings allowed Catholic church complexes to hold their own in the commercial downtown area. In the 1860s, St. Xavier's Academy expanded by building a four-story edifice on Wabash Avenue, adjacent to the Sisters of Mercy convent and St. Mary's Church; together, these buildings lined the Wabash Avenue block front between Madison and Monroe for a distance of over 150 feet, an area that none of the individual Protestant churches in the city commanded (see fig. 48). Similarly, on the north side, Holy Name Cathedral, St. Mary of the Lakes University, and related schools and residences filled an entire block (fig. 65). In parish after parish, schools, charitable institutions, and residences for members of religious orders provided expansive, architecturally harmonious settings for Catholic churches. Catholic churches were not as transient as Protestants churches were; substantial investment in religious precincts made it harder to move, and the influx of a German and Irish working-class constituency gave good reason to stay. The very changes that prompted the uprooting of Protestant congregations sometimes promoted Catholic rootedness. The scale and expansiveness of Catholic religious precincts (fig. 66) also helped these churches hold their own in the mixed residential, commercial, and industrial neighborhoods where many of the leading Catholic churches were constructed.

Attracted by Catholic monumentalism, Chicago Episcopalians made some efforts to the follow the Catholic model. As early as the 1850s, Bishop Henry John Whitehouse energetically campaigned for a *cathedral system* in the United States. Proposing that each bishop establish a cathedral church and complex, Whitehouse hoped that, as in Europe, bishop and clergy alike would "reside within the precincts of the cathedral" to form a prominent "ecclesiastical community." Whitehouse envisioned a parish training school, a theological department, and a chapter house and library connected with each cathedral, where free seating and daily services in different languages would welcome worshippers.[69] The plan bore striking similarity to England's Anglican cathedral system and to the institutional arrangement and landscape forms of the American Catholic church.

In planning his Chicago cathedral complex, Whitehouse drew strongly upon the architectural influences of the English ecclesiastical movement. In the early 1850s, Jacob Wrey Mould, a young English architect who had worked in Owen Jones's office and was familiar with the developing High Victorian Gothic revival, moved from England to New York.[70] In 1853 he made an unusual proposal to Whitehouse. Agreeing to furnish the design, working drawings, and specifications for the cathedral complex, Mould stated that if no other architect was consulted he would work "with no other prospect of remuneration than the being thoroughly identified with the great & blessed project and that you will seek as much as possible to bring my name in connection with your plan before the Christian & Protestant public of [your] vast community."[71] Promising his full "professional acquirements, energy, & zeal," Mould sought the Bishop's cooperation in "repressing all sham architecture & bad art in the Building project & . . . in carrying out everything in accordance with the spirit of the true Anglican church."[72]

Initial designs were drawn, but the cathedral project languished when Chicago's various Episcopal churches balked at the idea of financing a cathedral parish that might drain resources and members from established congregations.[73] This threat seemed pressing since Whitehouse had proposed locating the cathedral complex on land at Wabash Avenue and Jackson Street, the center of Chicago's most prominent elite residential neighborhood.[74] Cyrenius Beers, a member of the Episcopal church and owner of valuable adjacent residential real estate, had donated the land to the bishop.

Whitehouse finally established a seven-hundred-seat cathedral church in 1862 by purchasing an existing parish church. The Atonement Church, renamed the Cathedral of SS. Peter and Paul, had been built in 1854, designed in the English style of Gothic architecture by Theodore V. Wadskier and New York architect Henry Dudley (fig. 67). Atonement had fallen into debt during the economic crisis of the late 1850s, had moved from a pew rental system to a free-seating church, but still could not support itself in its poor neighborhood just west of the Chicago River. Whitehouse improved the church, giving it a new stone facade and a full, liturgically correct chancel, forming nine sides of a regular eighteen-sided

polygon. The ceiling of the chancel was painted in ultramarine with gold stars; the finely carved bishop's chair stood at the extreme back of the chancel flanked by twenty-four stalls for the clergy. Whitehouse did not reside near the church; however, a clergy house and a building for a Sunday school and an industrial school were later constructed. Even so, the complex fell far short of Whitehouse's grand vision and in terms of scale and form was surpassed by many local Catholic parish complexes.[75]

With denominational support, however, the bishop's church stayed rooted in a poor, unfashionable urban neighborhood. Protestant "institutional churches" established in poorer neighborhoods at the end of the nineteenth century united churches with schools, gymnasiums, meeting facilities, settlement houses, and libraries, similarly drawing on the Catholic model. These examples were clear exceptions, however. The leading tendency of nineteenth-century Protestant religiosity in Chicago was toward mobility.

Both aspects of the reconstructed Protestant landscape—new church siting and Gothic monumentality—came under criticism from some contemporaries. The prospect of a churchless downtown and of abandoned houses of worship sparked particular concern. The insulation and monumentality achieved by "seeking eligible corner-lots where the better classes dwell" was tempered by the prospect of "Gothic Arches . . . going the way of all earth!"[76] Churches destroyed to make room for stores or tenements or, worse yet, converted into boarding houses, dance halls, saloons, or exhibition places for "cheap statuary" suggested a lack of reverence for consecrated ground and existing religious structures.[77] In 1867, the Grace Episcopal Church vestry balked at a proposal to sell its Wabash Avenue building for a commercial development. Instead it sold to a leading Jewish congregation, Kehilath Anshe Mayriv.[78] Yet in 1885 religious critics reported that all too often "the plat of land remains 'holy ground' only as long as it is a convenient building site. . . . If a congregation has an advantageous offer for their church edifice they generally sell it without compunction or regret."[79]

In 1869, Bishop Whitehouse introduced a new church canon to "arrest the sacrilege" of destroying churches and disposing of land for "unhallowed, worldly and common use, merely for the luxury or convenience of pew owners, where God clearly indicates a righteous claim to His own."[80] Whitehouse objected to the impatience for removal, the "chase after wealth, in which city churches, alas, are such Nimrods."[81] By requiring the

64. Holy Name Cathedral interior, Patrick C. Keely, architect, 1874 (photo ca. 1885). Chicago Historical Society, ICHi 19192.

65. Holy Name Cathedral, Patrick C. Keely, architect, 1874
(photo ca. 1885). Building to the left is the Academy of the
Sacred Heart. Chicago Historical Society, ICHi 22305.

66. Church of the Holy Family, Dillenburg and Zucher, architects, 1857–60 (1898 photo). To the right of the church is St. Ignatius College (1869); at the rear is the Convent of the Holy Heart of Mary. Avery Library, Columbia University.

67. Cathedral Church Of SS. Peter and Paul, Henry Dudley and Theodore V. Wadskier, architects, 1854 (photo ca. 1870). Chicago Historical Society, ICHi 22318.

bishop's approval for relocation, the diocese hoped to preserve consecrated lands as well as to foster a more orderly network of urban parish churches.

The new Episcopal canon stemmed congregational interest in abandoning the building Trinity Church had occupied for eight years on Jackson Boulevard. Designed with twin towers and domes in 1861 by Wadskier, the church had failed to provide the focus for a stable parish. In 1867, the Trinity rector reported the loss of several "influential members" to newer neighborhoods on the city's south and west sides. This "drift of population" would "ere long be seriously felt by Trinity church."[82] Convinced that the "necessities of business would drive Trinity Congregation from its . . . locality," the vestry began searching in 1869 for a site between Eighteenth and Thirtieth streets. Members of the congregation who objected to the prospect of the church becoming a warehouse or music hall suggested that, in case of relocation, the building should remain open as a free church supported by the city's various Episcopal parishes.[83] This idea implied that despite business expansion many potential worshippers would be left in the area around the church. After Whitehouse's criticism of church transience, the Trinity rector insisted that business encroachment increased the need to stay put and to gather in the floating population: "Should the church beat a hasty retreat before mammon? Should the central portion of the city be wholly given up to mammon, as was Athens, in ancient times, to idolatry?"[84] Emerging from the most centralized of the Protestant denominations, the negative answers to these rhetorical queries forcefully challenged the city's new religious landscape.

In the wake of the dislocations caused by the 1871 fire, individual congregations and entire denominations again debated the propriety of religious location. The fire destroyed much of the settled portion of Chicago. It swept across nearly three and one-half square miles, destroying churches from St. Paul's and the First Presbyterian on the south side, through the Second Presbyterian at the center, through St. James and the Cathedral of the Holy Name on the north side. In all, thirty-nine churches valued at $2,281,500 burned. Churches that had shared in the city's prosperity now shared equally in its ruin. Many congregations returned to the makeshift settings characteristic of early religious worship in the city. The fire provided the opportunity for restructuring land use in the center of the Chicago. It is conceivable that concerted action by Chicago's various denominations could have led to the reestablishment of downtown religious precincts protected from commercial intrusion. In reality, just the opposite happened. The postfire rebuilding accelerated the 1850s and 1860s' trend toward a specialized commercial center and a decentralized religious landscape.

Reviewing the postfire building of the city in 1880, a Chicago Baptist newspaper called the *Standard* bemoaned the continuing abandonment of the "Heart of the City" by Christian churches. It insisted that the drift of population away from the center after the fire had been overestimated, while the need for downtown churches and religious institutions had been underestimated or, worse, not even considered. With the exception of a few missions and churches, "evangelical influence has left the heart of the city and gone to the extremities." Challenging this pattern, the *Standard* asserted, "Where the world is most, there is the place for religion, and that is always 'in the heart of the city' . . . by making the worldly center in each such community the strong-hold of gospel teaching and gospel power, it may thus be practically shown that Christians do not look upon their religion as suburban and ornamental merely."[85] Regardless of such appeals, individual congregations continued to cast their lot with more promising "fields," the outlying residential neighborhood. Congregational transience left in its wake abandoned sacred ground, commercially violated religious precincts, and a human flock tended only by distant shepherds.

As architects and congregations collaborated to give the churches in Chicago's outlying neighborhoods an increasingly monumental form, some observers raised their voices to protest the grandeur of the religious landscape. Such critics made careful rhetorical distinctions between a *church,* meaning a congregation of people, and a *church building, church edifice,* or *church structure.*[86] Poorer religious societies took comfort in their understanding that it made "no difference what kind of house they worshipped in, if they did it sincerely . . . God would bless their efforts and crown them with success."[87] In 1877, sadly celebrating their last service before foreclosure and eviction by the mortgage holder, the St. John Church rector insisted that, "The real parish life is not in the beautiful buildings that it may rear . . . not in the external ritual that is garnished with eloquent oratory and fine music. . . . The real life is within."[88] This view comforted many Chicago congregations that attempted to preserve church life after the 1871 fire destroyed their churches. It depended on an ability to disconnect the link between external and internal that had long informed Protestant religiosity in America.[89]

Having severed the link between the material and the spiritual, critics saw no need for churches to surpass the grandeur of secular forms in a society already considered too materialistic. Grand architecture encouraged "the worship of the material element in religious life,"[90] and even worse, it substituted "worship of churches for church worship."[91] Money spent on expensive churches deprived the poor of religious and social services and often hampered religious spirit and congregational harmony.[92] In 1877, the mortgage debt on Chicago's Protestant churches exceeded one and a quarter million dollars, enough money to build dozens of more modest churches.[93] Debt distracted religious leaders and church members from spiritual matters and even hastened the departure of old members while keeping away new ones. Thus the monumental projects could effect the very opposite of what church builders sought.[94] In the late 1860s, St. John's Episcopal Church, for example, had set out to build a church that would "excel in size and beauty and magnificence" all other churches on the west side.[95] As debts piled up and economic depression slowed the project, congregational and pastoral transience ensued. Upon resigning, one rector at St. John's lamented the "mistake" of planning so grandly; owing to debts and an unfinished building, "Rumors prejudicial to the interests of the Parish were set afloat and extensively credited. Families of Episcopalians coming to our neighborhood were generally deterred from connecting themselves with the church through the fear of incurring an uncomfortable burden, while old members dropped off to escape what they thought was inevitable."[96]

The advocates of modesty in church design enthusiastically pointed to the "magnetic" and sympathetic religious services of congregations forced into cramped quarters after the Chicago fire. Observing plans to rebuild, "to return to the flesh-pots of Egypt and gilded temples of the old time regime," critics wondered somewhat mournfully, "Were it not well for the soul to always live in plain houses, for the soul's culture to be always so paramount in importance as to make it indifferent to the material surroundings of that culture?"[97] Although certain Protestant groups had long advocated aesthetic restraint and warned against sumptuous "Catholic" forms, antimonumentality in the nineteenth century was influenced less by religious tradition than by critiques of powerful secular interests apparent in church building.

Proponents of monumental ecclesiastical architecture countered that church design should reflect Chicago's progress, prosperity, and religious commitments. They envisioned a close symbolic link between the church as congregation and the church as building. They gave little credence to the idea that grand buildings necessarily masked spiritual depravity. For Rev. William

Everts, who led monumental building campaigns at the First Baptist Church in the 1860s and again in the 1870s, churches should be the leading buildings in a community and "a fair representative of the state of culture and wealth which surrounds [them]."[98] In a book on public worship illustrated with designs by W. W. Boyington and James Renwick, Everts declared that a cheap church in a wealthy city attested to a "poverty of religious sentiment." He concluded, "Parsimony and avarice in a people are far more to be dreaded than extravagance in church building."[99] Thus, advocates of monumental religious architecture turned on its head the critique of ostentation and materialism; they insisted that grand church buildings actually provided evidence that pervasive commercialism had given way to selfless expressions of urban beauty and spiritual life.[100] In dedicating the Union Park Congregational Church, perhaps the finest prefire church, Rev. Charles D. Helmer strongly defended grand church architecture. He argued that "the expression of the ripest culture" of the age meant fostering tradition, and not the adherence to modern utilitarian demands that all buildings be merely "serviceable only for use and shelter." Such utility, according to Helmer, would obliterate the crucial distinction between churches and more worldly structures such as barns and factories. When asked how congregations could build expensive churches when so many poor people needed relief and mission work, Reverend Helmer rejoined: "If churches in the city had no distinctive architecture, a great moral impression would be lost. . . . If they built plain churches, the world's people would say the Christians cared little for their God, for they built Him such poor houses of worship, while their private houses were sumptuous. . . . Moreover, nothing was too good for God."[101]

Here, in complex form, was the fundamental commitment to and entrapment by materialism of the city's middle and upper classes: to an unprecedented degree, America's middle class defined itself in terms of its possessions, appearances, and built environment. As the nineteenth century progressed, the consumption of numerous tastefully embellished consumer items became increasingly important to middle-class Chicagoans. Houses were, as Reverend Helmer had suggested, necessarily sumptuous. Located nearby, churches could not stand the comparison unless they too were sumptuous. In his concern for the moral impression created by churches, Helmer expressed the common assumption that external appearances would have profound impact on internal states. Helmer and his allies were people who

felt the press of their environment strongly; they took their identity from things.

Nonetheless, a few exceptional congregations abandoned prevailing forms of religious architecture in order to preserve traditions of the centrally located church, and their experience suggests the strength of countervailing currents in nineteenth-century culture and religiosity. Of the six leading Washington Street churches in the 1840s only one, the First Methodist, showed any determination to remain in place during the 1850s and 1860s. As its erstwhile neighbors moved to Wabash and Michigan avenues, the First Methodist congregation built a strikingly novel hybrid building that combined religious worship with commercial space. The decision to give up a distinct specialized building in order to remain downtown was not fully voluntary. Unlike neighboring congregations, the First Methodist could not sell its Washington Street lot to raise a monumental building fund. The 1839 deed stipulated that if the congregation moved, the land would revert to the heirs of the original owners. Yet the pastor at First Methodist made a virtue of necessity, thanking God that the church was "anchored fast upon that corner, in the very center of business."[102]

The First Methodist congregation erected its "novelty among church edifices" in 1857, when the improvement of Washington Street left its twelve-year-old Greek revival church nearly five feet below grade. Rather than raising the building to grade, the congregation commissioned Edward Burling to design a new one. Burling demonstrated his abilities as a Gothic revivalist in his contemporaneous designs for St. James and other Chicago churches.[103] For the first Methodist, however, he designed a four-story structure, 80 feet by 130 feet, of smooth Athens limestone. The building had the exterior appearance of a "first class business structure."[104] Eight stores and an arched entrance to the second floor offices lined the first-floor facade; the church occupied the third and fourth floors (fig. 68). Inside, Burling introduced a level of grandeur to the church's auditorium by making the 62-foot by 84-foot room extend through both the third and fourth floor—giving its ceiling a height of 33 feet. The auditorium's amphitheatrical seating plan, probably the first in a Chicago church, accommodated a thousand people. The plane ascending from the pulpit gave everyone a "full and unobstructed view of the speaker."[105] Ornate woodwork and fresco designs embellished the religious interior.

First Methodist's officials used the building's rental income to compensate for the loss of support from members who had moved to the Wabash Avenue Methodist

68. First Methodist Episcopal Church block, Washington Street, Edward Burling, architect, 1857 (photo ca. 1870). Chicago Historical Society, ICHi 22330.

and other outlying churches. The rental income was substantial, an estimated $15,000 a year, and it supported First Methodist and provided other Chicago Methodist churches with $725,000 over the next fifty years.[106] The First Methodist provided for religion at the center of Chicago's commercial downtown and at the same time viewed its building as an asylum. The church's pastor insisted that the "necessary" commercial floors effectively lifted "the place of worship high above the noise and din of the street."[107] First Methodist laid claim to a successful, alternative religiosity—one that combined removal from some downtown distractions with a forthright attitude toward money making.

The 1871 Chicago fire destroyed the First Methodist Church block, and the congregation built a new hybrid commercial and religious building (fig. 69). In the early 1890s, continuing to occupy one of the most valuable corners in downtown Chicago, the First Methodist congregation made plans to replace its postfire office block with a skyscraper.[108] In the 1880s and 1890s, construction of skyscraper office buildings by Chicago religious, fraternal, and cultural organizations, such as the Auditorium, Masonic Temple, and Women's Christian Temperance buildings, offered models for the First Methodist's transition from low-rise structure to skyscraper. The depression of the 1890s delayed these plans, which finally culminated in 1924 with a twenty-two-story building. Designed by Holabird and Roche, the building contained the church and eighteen stories of commercial offices. The French Gothic design, complete with tower, spire, and buttresses, introduced religious expression that had been omitted in the two earlier Methodist Church block designs (fig. 70). In fact, with a 568-foot spire and with some liberties taken, the structure was designated the tallest *church* in the world.[109] For most of the late nineteenth century, however, the Methodists had not felt the need for a distinctively religious style.

In Chicago, structures that combined commercial and religious uses were often referred to as having the "Methodist Church Block Style." In the 1870s, several churches built and occupied religious auditoriums planned in connection with stores and offices.[110] In 1869, William Bross proposed that his Second Presbyterian congregation build "the best business block that can be constructed" and place an auditorium for the church's services on the second floor. In 1871, in the face of growing interest in relocation, Bross proposed that the congregation take the fixtures, organ, and name of the church and move south of 12th Street to a location "not so near any other Presbyterian church as to materially

69. First Methodist Church block, Washington Street, architect
unknown, ca. 1874 (photo ca. 1894–10). Chicago Historical
Society, ICHI 19985.

70. Chicago Temple Building/First Methodist Church, Washington Street, Holabird and Roche, architects, completed 1924
(photo ca. 1925). Chicago Historical Society, ICHI 22300.

affect its prosperity or welfare." Selling the church property for $160,000 would fund a splendid new church. However, Bross questioned whether such a move would be in the spirit of the congregation's founders, William H. Brown, John C. Williams, and Benjamin Raymond. He insisted that they would all prefer his plan to leave the property in trust for all the Presbyterian churches in Chicago, improve it with a substantial business building, and turn the second floor over to a centrally located "Union Presbyterian Church" and the offices of various church organizations. By a close vote, the trustees defeated Bross's proposal and the Second Presbyterian sold out and moved south, taking its real estate profits along.[111] In the early 1890s, Chicago's Episcopal diocese planned to build a downtown skyscraper. The skyscraper, "the grand enterprise which makes illustrious the age in which we live," would give the church prominence, institutional space, and substantial rental income in the center of the city. The church abandoned the project during the economic crisis of the 1890s.[112]

Hybrid commercial-religious structures and plans constituted an architectural response to the broader debates over religion's physical place in the city. The few congregations that were determined to stay downtown tended to permit business forms to encroach substantially upon the architectural symbolism of the religious landscape. The hybrid building recalled the earliest worship services in the city, which had taken place in hotels, saloons, and auction houses. The Methodist church block style challenged elaborate expressions of Gothic revival religious architecture and contributed to debates over the relevance of monumentality to the religious life of the modern city.

As Chicago's leading congregations conservatively commissioned Gothic revival churches for outlying residential sections, the city's two most prominent religious modernists, David Swing and Hiram Washington Thomas, overthrew the prevailing forms of the religious landscape.[113] Heresy trials in the 1870s and 1880s prompted both men to leave their Gothic revival style churches and their denominations. In establishing independent congregations, they reclaimed the city center as the place for religion. Stylistically, their new churches undercut prevailing wisdom concerning the need for architecturally distinct buildings for worship. Their congregations thrived on the social heterogeneity and dynamic experience of the city that prompted the transience of most other congregations.

In 1866, after several years as professor of classics at Ohio's Miami University, David Swing took the pulpit at Chicago's Westminster (Fourth Presbyterian) Church. A mild, soft-spoken preacher, Swing attracted a large congregation by delivering compelling sermons on religion and modern life. Swing espoused a decidedly liberal theology. He spoke less about Presbyterian creed than about the daily concerns of his congregation, while at the same time expressing faith that God was manifested in life and in the approaching Kingdom of Heaven on Earth. A contemporary account insisted that those expecting the "usual orthodox discourses will be sadly disappointed . . . [for] he speaks in a new way, and draws his illustrations from the resources of a large and liberal reading, and from the manifold instances of nature, science, and art."[114]

In 1874, Francis L. Patton, a professor at Chicago's Seminary of the Northwest, charged David Swing with heresy. Trained at the Princeton Theological Seminary and chosen for his Chicago position by the theologically conservative industrialist Cyrus H. McCormick, Patton adhered to old-school Presbyterianism. Patton charged that Swing's sermons failed to maintain gospel truths and church creed. In the most famous American heresy trial of the decade, the Chicago Presbytery found Swing innocent of Patton's charges. When Patton moved to appeal the case to the national Presbyterian council, Swing resigned his pastorate to avoid further prosecution and promptly established an independent church in Chicago's downtown.[115]

Swing's new site was McVicker's Theatre, where his Presbyterian congregation had temporarily held services after the 1871 Chicago fire. Swing's comfort at settling into a theater for religious worship reflected his sanguine view that theater and religion both represented benign, cultured possibilities of city life. Dismissing country life as unequal to society's fullest potential, Swing declared that "the word 'city' embodies so much of all that is great and beautiful in life that I shall consider it as a term embracing all human achievements and qualities of greatest worth. In the world's cities have always been gathered the arts, the sciences, the religions, the philosophies, the education, and refinement of the race."[116] Swing was considered the "high priest of the beautiful" and the refined in the midst of a city which, he admitted, was thought to "appreciate the art of wrestling . . . butchering swine, and . . . fixing election returns more than Lessing's 'Lacoon' and the 'Seven Lamps of Architecture.'"[117] Swing felt that Chicago contained two lives, a commercial life and a nascent and vital intellectual life.[118]

Swing's use of a theater for his Sunday services underscored his own preference for modern theology and for modest church building. He preached that Christianity always bore the imprint of the age. He argued that Chicago's specialization in life's necessities—meats, grains, lumber, and iron manufactures—found a counterpart in the local interest in "fabricating a new form of Christianity" pervaded by a "strong practical sense" that dismissed the "abstract themes" of Calvinism and ancient theology.[119] Swing's "strong practical sense" of religion led him to deplore what he called "limestone Christianity." He appreciated aesthetic pursuits and grand architecture; yet, for modern religious practice he declared that "the Christian world has not money enough to carry forward its great moral reform, and at the same time gratify a great architectural passion. It must from a simple want of money abandon one or the other." The urban masses needed plain churches, Sunday schools, libraries, and lecture rooms "planted in the centres of their daily life" and not the moral education of "cathedrals built upon fashionable streets." Swing pleaded for "an age of the useful, an age of plain churches for the people and plain theology, too, for the pulpit and the people alike."[120]

Swing's religious utilitarianism narrowed, or rendered irrelevant, the long-standing distinction between secular and religious pursuits that had pervaded Chicago's landscape. In fact, Swing's Central Church annexed the theatrical forms of popular culture while adopting the framework of existing cultural undertakings. Wealthy community leaders, many of whom belonged to other congregations, financially supported the founding of Swing's church. Drawn perhaps by Swing's liberalism and cultivation, leading city figures such as Frank Blair, William Bross, Wirt Dexter, John B. Drake, N. K. Fairbank, Franklin MacVeagh, John A. McVicker, Joseph Medill, Ferdinand W. Peck, H. H. Wilmarth, and forty other Chicagoans pledged a thousand dollars each to cover debts in Swing's church. Debts never developed. In 1875, when annual seats went on sale for the Central Church, ten thousand dollars' worth of seats sold on the first day. Carpenter and Sheldon, theatrical agents and organizers of popular entertainments, managed the ticket sales. Annual tickets cost from $12 to $20 for the parquet and dress circles, $8 to $10 for first balcony, and $5 for second balcony.

Seats in Swing's church cost only a fraction of the pew rental fees collected in other churches. Supporting, among other things, monumental building projects, the pew rental and assessment system provided the primary system of support in Chicago's nineteenth-century Protestant churches. Starting with rows one-quarter or one-third of the way back from the pulpit, the highest priced

pews stood in the middle of the congregation. When members occupied their pews, they filled in a clearly perceived, settled hierarchy based on financial support for the church (fig. 71). Within the congregation every family had its place. Accommodating a fairly narrow range of income levels, the pew rental system tended to encourage social homogeneity by excluding the poor, the outsider, and the less committed worshipper. In the midst of the transient social conditions of the city the pew rental system provided a degree of unambiguous order.

Swing's lower prices allowed for the congregation to be substantially more diverse. One hundred people lined up out the Jansen and McClurg booksellers shop even before tickets went on sale for places in Swing's church. In the crowd stood people ranging from the "thorough churchman" to the "pensive young man," to the "woman out shopping," to the "shrewd speculator," hoping for a profit from ticket resales. In short, "every diversity of character was represented."[121] The congregation that actually gathered for services was often even more diverse than this would suggest: over half the seats in Central Church were left open to be occupied for free. Swing could "revolutionize" church practice, said one newspaper, and it pointed out his debt to popular urban culture: "Chicago already has popular Sunday lectures on science, art, and other useful topics; Prof. Swing's Sabbath addresses on religious themes, delivered in a theatre, with the same surroundings and much of the methods of popular lectures, and delivered to popular audiences . . . is an innovation."[122]

From McVicker's Theatre, Swing challenged the prevailing forms of "limestone Christianity" and cast aside the comforting images of settled congregational stability for a more cosmopolitan ideal of community. "Not much should be said of the fellowship and friendship that sprung up in the regular house of God," he insisted, for the formality and exclusiveness of the "congregations on the avenues" prevented fellowship. He proposed instead a congregation transcending neighborhood associations and open to the diversity of city life. People would easily travel downtown from outlying areas on the same transport that carried them to work, to shop, or to attend commercial entertainments. In outlining his vision of Central Church, Swing emphasized that "each city is full of strangers. We live each door to each door and live unknowing and unknown. Here, where you will have regular seats, and where some of the stiffness of the more formal churches will be wanting, you will soon reach an acquaintance with your neighbor, and a final knowledge of all not to be found in churches which would seem to

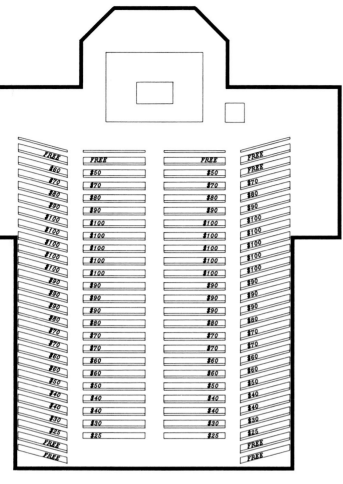

71. Pew rental plan, Trinity Church, 1877. Schematic drawing by Areta Pawlynsky, based on original plan.

promise more."[123] In its first year, Central Church began drawing some of the largest Protestant assemblies in the city; the worshippers appreciated services "without the oppressive tax . . . of [the] exorbitant pew-rental system." Swing invited Chicagoans to occupy rented or free seats, "without regard to sect, social consideration, or any distinction."[124] Swing's success revealed limitations of the conception of a traditional, rooted congregational life.

In January 1880, Swing's Central Church moved into the Central Music Hall (fig. 72), a new building designed by Dankmar Adler and Company and built with the understanding that the Central Church would become a main tenant. George B. Carpenter, a theatrical agent and member of Swing's congregation, promoted the building. He enthusiastically endorsed both the hybrid commercial-religious plan and the secular, commercial use of church auditoriums. He criticized other Protestant congregations for their inability to "escape the influence of the architectural symbols, of the religious faith of the fifteenth century." He condemned the "foolish clinging to ancient symbols" and "absurd extravagance of $150,000 to $200,000 capital idly invested in a structure that is in use six hours per week."[125] Central Music Hall embodied Carpenter's ideals; six stories of stores and offices fronted the grand music hall at the corner of State and Randolph streets. The auditorium, with its superior acoustics, provided a good revenue when not in use for Central Church. That the building served Mammon so extensively meant that church expenses were modest. In 1880, 1,004 members paid $5 to $25 for seats, and 1,500 other seats were filled with nonmembers at no cost (fig. 73). The church's success continued until Swing's death in 1894.

Although never as popular as David Swing, Hiram Washington Thomas and his People's Church congregation followed Swing's example. After moving to Chicago in 1869, Thomas became the pastor of the Park Avenue Methodist Church. Three years later he transferred to the First Methodist Church where, from the church's hybrid commercial-religious structure, he preached in the face of growing residential transience. In 1873, he ran afoul of theological conservatives when he helped found the Chicago Philosophical Society. He was assigned to a less prominent position in Aurora, Illinois, but returned after two years to head Chicago's Centenary Methodist Episcopal Church. In 1880, church officials asked Thomas to resign because of his alleged lack of adherence to Methodist doctrine, in particular his divergence on the issues of inspiration, atonement, and future punishment. Convicted of heresy and expelled from the

72. David Swing's Central Church, Central Music Hall,
Dankmar Adler and Company, architects, 1879 (1888 photo).
Chicago Historical Society, ICHi 22298.

73. David Swing's Central Church auditorium, Central Music
Hall, Dankmar Adler and Company, architects, 1879 (photo ca.
1880). Chicago Historical Society.

church in 1881, Thomas organized the independent People's Church.[126]

As in the case of the Central Church, private parties guaranteed the People's Church against debt; twenty-seven supporters each pledged $250. The congregation moved to Hooley's Theatre, filling all the seats and aisles. Thomas preached that "all places are holy when used for holy purposes"; a simple cabinet organ, a chair, a table, and a bouquet of flowers transformed the stage and theater into a church. According to Thomas, at the People's Church, "a large personal liberty in matters of belief would not be a bar to admission nor a cause of complaint or criticism."[127] Material limitations would not debar worshippers either: seat rents ranged from $5 to $10 a year, and there were no special additional collections; free seats provided for visitors and the poor. The People's Church would know "nothing of nationalities or conditions of men."[128]

Swing and Thomas countered the nineteenth-century dislodging of religion from the physical and cultural center of American urban life. They succeeded in part because of their willingness to collapse physical and theological distinctions between secular and religious forms. Where other church leaders were willing to adopt modern amphitheatrical interiors, Swing and Thomas accepted the entirely secular settings of Central and Hooley's theaters for congregational meetings. To critics of Swing's ministry, his use of Central Music Hall seemed emblematic of the final succumbing of religion to commercial encroachment. For their part, Swing and his supporters viewed the bastions of "limestone Christianity" as the retreat of organized religion to a marginal position in the face of a complex, modern city. The social and economic heterogeneity and cosmopolitanism of downtown, which drove so many congregations to outlying neighborhoods, proved attractive and compelling features of the Central and People's churches. Many other congregations appeared buffeted by the scale and pattern of urban change, while Swing and Thomas's congregations cast their worship in a distinctly modern form, taking advantage of the structure of modern metropolitanism. Regardless of the place people occupied on the spectrum between Swing and the fashionable churches, it was clear that attitudes toward commerce and views of the competition between religion and popular culture played a central role in defining religious form.

The heretics did represent the exception in Chicago. Movement away from the center, begun in the 1850s, continued into the twentieth century. The link between home and church, first explored at midcentury, endured. Expanding businesses continued to uproot congregations from their neighborhoods. Increasingly, however, changes in residential population rather than in commercial building prompted congregational transience. In 1895, for example, the Trustees of the First Presbyterian Church declared that its surrounding area was not attracting a "stable resident population." A few years later, after watching the "large number of wealthy families" who had lived near the church move away, the trustees lamented that "the pew renting material in this vicinity is becoming more limited."[129] Churches dominated outlying residential areas as well as an emerging suburban landscape. Editors of the *Interior* observed in 1892 that many people, "seeking relief from the crowded city" had moved to "this girdle of hope, the broad green belt of suburban residence."[130] Although they did not often match the wealth and grandeur of leading city neighborhood churches, suburban churches stood out prominently in the small commercial and civic centers planned around commuter train stations. In the suburban towns many churches regained a position in the center of the community, a position they had yielded in the larger city.

A world of difference separated Swing's response to the modern city from the majority view, represented by neighborhood and, later, by suburban churches. When people emerged from leading houses of worship in the 1840s and 1850s, they found themselves amid the social, economic, and architectural diversity of the city. When they emerged from their churches after the 1860s and 1870s, they often faced elite residences, secluded from many characteristic elements of commercial Chicago. This reorientation expressed broad changes in organized religious life. In the seventeenth and eighteenth centuries, state-supported churches demanded the attendance of residents and stood as the central focus of the American city. In the course of the nineteenth century, voluntary church congregations and growing secular diversions from religious engagement were palpably manifested by the ceding of central ground to business activity and to secular, commercial entertainment. Church builders and congregations thus sought new settings in which to express commitments to religious life.

Notably, though, Swing and Thomas, unlike earlier church and park builders, did not view downtown as inimical to their missions. They recognized the emerging power of downtown to participate in the images and expressions of spirit, refinement, and uplift even in the midst of business. These juxtapositions of apparently disparate social ideals grew increasingly important as Chicago city builders and architects shaped the modern commercial skyscraper.

Chapter 4

"A City under One Roof"

Skyscrapers, 1880–1895

Woman's Temple Building,
southwest corner of La Salle
and Monroe, Burnham and
Root, architects, 1890–92.
Postcard (ca. 1900) from
author's collection.

In the 1890s, Chicago's thirteen-story Chamber of Commerce building easily impressed downtown visitors. One guidebook declared it "in many respects the finest commercial structure in the world and certainly one of the grandest office buildings in the United States."[1] Although built of steel and terra-cotta, with "as little wood as possible," the building would not have gone up without lumber.[2] For over thirty-five years the building's owners, Perry Hannah, Albert Tracy Lay, and James and William Morgan, had been partners in the lumber business. Starting in 1850 with six thousand dollars, they had made a fortune by extracting pine and hardwood trees from more than forty-five thousand acres of Michigan forest. They ran the trees through their mills at Traverse City and Long Lake and shipped them on their steamers to Chicago, the world's largest lumber market. Then in 1886 they sold their depleted lands and their mills for over three-quarters of a million dollars.[3] Investing $1½ million in the Chamber of Commerce project, they sought a new fortune from real estate (fig. 74).

The Chamber of Commerce and other skyscrapers dominated Chicago's downtown in the 1890s (fig. 75). In these tall buildings, a host of economic, technological, and aesthetic factors converged, adding a new dimension to the cityscape. Skyscrapers literally and figuratively overshadowed other urban buildings, landscapes, and institutions. Men like Perry Hannah were accustomed to such dominance. Hannah had developed Traverse City, Michigan, a town of six thousand in 1886, from the wilderness surrounding his company's lumber operation. In Traverse City, Hannah's mill dominated the waterfront, his mercantile business ("Dealers in Everything, Universal Suppliers") dominated downtown, and his forty-room mansion, built in the 1890s, dominated the town's elite residential neighborhood.[4] In Chicago, skyscraper buildings effected a new dominance. Although it incorporated the lower walls and entrance portico of the 1872 Chamber of Commerce, Hannah's building burst the orderly relation that had existed between that earlier structure and the prominent city hall and county courthouse across Washington Street. The new Chamber of Commerce building now cast an afternoon shadow on the courthouse.

Even before skyscrapers took possession of the downtown cityscape, Chicagoans had worried about the scale of commercial buildings there. The low-rise warehouses, wholesale businesses, and other commercial buildings of the 1850s and 1860s had seemed large enough to "submerge" downtown churches and residences. Now high-rise monuments to commerce towered over other city structures, frankly proclaiming business as the heart of

the city. When they commissioned skyscraper office buildings, Chicago's city builders were not creating a sentimental realm for residence, religion, the contemplation of nature or gregarious promenading. They were operating out of their own pockets and for the sake of profit, multiplying rental income from a single urban lot many times over by piling up floor space with the aid of modern building technology. By their height, expense, and status as complex tools for money making, skyscrapers were expressive of the city's prosperity, competitiveness, and the aggressive pursuit of private goals. Here, surely, in these "hives of business," was the home of Mammon, unfettered and unalloyed.[5]

Indeed, historians have interpreted skyscraper technology and aesthetics as direct outgrowths of Chicagoans' pragmatic, money-making goals. To many, skyscrapers represent the triumph of function and utility over sentiment—and, more fancifully, the triumph of America over Europe or, perhaps, the frontier over the civilized East. In this view, these new, tall buildings expressed a forthright acceptance of new technologies and (equally importantly) an aesthetic that prized an honest declaration of structure. The weight of the historiography surrounding the "Chicago School" or "Commercial Style" presents the skyscraper as a triumph of stylistic modernism. This interpretation has been so dominant that it merits scrutiny of its premises, conclusions, and those myths it has fostered.[6]

Sigfried Giedion and Carl W. Condit pioneered serious architectural history of the Chicago skyscraper in the 1940s and 1950s.[7] They shared a critical enthusiasm for twentieth-century European and American modern architecture. Both lacked sympathy for aesthetic and cultural expressions that were embodied in eclectic design. For them, sentimental suggestions of past cultures and styles amounted to a banal evasion of artistic responsibility. From this viewpoint, the Gothic facades of churches in Chicago's residential neighborhoods seemed shallow and ungenuine. Artists and architects should express their own age with their own styles. Condit wrote: "The artistic failure of architecture in the nineteenth century can be stated very simply: It was the failure to form a style. It was the failure to provide, in its own vocabulary, an aesthetic discipline and an aesthetic expression of science, technology, mechanized industry, and modern urban life."[8] Here Condit wove together two premises, one about the nature of architecture, the other about the nature of nineteenth-century urban society and culture. Both Giedion and Condit sought in the past and proposed for the present a unified art and culture, and therefore a style of building that would be the physical

74. Chamber of Commerce Building, southeast corner of Washington and Lasalle, Baumann and Huehl, architects, 1888–90 (photo ca. 1895). The tall building at the right is the Tacoma Building, northeast corner of La Salle and Madison, Holabird and Roche, architects, 1887–89. Chicago Historical Society, ICHi 00283.

75. Chicago's downtown skyline (1900 photo). Skyscrapers of
the 1880s and 1890s rose above the postfire skyline of walk-up
commercial buildings. The tall buildings include (from left to
right): in the distance, the Ashland Block; in the middleground,
the rear wall of the Title and Trust Building, in the background,
the top of the Unity Building, the Masonic Temple, in the foreground, the rear wall of the Hartford Building. Library of
Congress.

representation of Zeitgeist. Identifying the *spirit* of the nineteenth century as industry, technology, and commerce, they reasoned that these forces required symbolic codification. The challenge was for the architect to realize a new style based on the new materials, technologies, and forms of modern building.

More critical than historical, these analyses of Chicago skyscrapers were not entirely lucid about the factors that prevented designers from easily meeting that challenge. As Giedion formulated the historical process, an inherent logic unfolded within *architecture* and *construction* such that these abstractions acted nearly as historical agents, with lives or motivations of their own. "For a hundred years architecture lay smothered in a dead, eclectic atmosphere in spite of its continual attempts to escape. All that while construction played the part of architecture's subconsciousness, contained things which it prophesied and half revealed long before they could become realities."[9] With architecture pressing forward, what held architects back was a fatal ambivalence. *Cultural schizophrenia* or a *split personality* beset nineteenth-century Americans, and a *split civilization* compromised nineteenth-century art and architecture. A *schism* or gap existed between architecture and engineering, design and construction, art and industry, feeling and rationality.[10]

Deliverance from dead eclecticism, so the account goes, came when architects began to draw stylistic inspiration from the new building techniques—and in particular from the steel-frame structural system. Chicago's 1880s commercial architecture succeeded in bridging the gap between "bare construction" and "architecture in the grand manner." Giedion concludes: "With surprising boldness, the Chicago School strove to break through to pure forms, which would unite construction and architecture in an identical expression."[11] Thus skyscraper architects confronted the "technical and aesthetic problem of creating in masonry a form appropriate to the needs and the spirit of the new commercial and industrial culture." They created a stylistic "revolution." They fostered an aesthetic "emancipated from the last vestige of dependence on the past."[12] In short, early skyscraper designs created by Chicago architects prefigured important elements of twentieth-century modern style and form.

Seen from modernism back, then, Chicago skyscrapers might be celebrated as unalloyed expressions of modern, commercial life. Neither Condit nor Giedion detected reservations about capitalist society in Chicago's tall buildings. Framed in this manner, Chicago's spectacular skyscrapers would seem to provide evidence that whatever concerns about commerce had earlier informed designs for Gothic revival churches and the movement for public parks, these had been resolved by the 1880s—if not by city builders as a whole, then at least by those individuals who built skyscrapers. Skyscrapers might establish the maturity of bourgeois culture in the last decades of the century—the culture's reconciliation with material striving and the healing of the schizophrenia of the times.

There is little doubt that skyscraper clients and architects accepted—and, indeed, centrally participated in—commercial and industrial capital. Like Giedion and Condit, late nineteenth-century architects, builders, and critics of Chicago's skyscrapers felt that commerce, industry, and technology largely defined the spirit of the city. Nonetheless, both in word and in stone, the Chicagoans of the era demonstrated social and cultural ideals distinct from those attributed to them by their historians.

This analysis of Chicago skyscrapers draws on but modifies Giedion's and Condit's view. It does not begin with the rise of the skyscrapers' enabling technology—steel-frame construction, elevators, terra-cotta fireproofing, or novel foundation techniques—or with the aesthetic challenge presented by unprecedented building heights. It begins with aspects of the skyscraper that impressed contemporaries rather than what impressed modernist architects and critics of a later era. From this view, it appears that Chicago architects made a greater effort to transcend commercial utility than to express it symbolically. The reconciliation of city builders with commerce did not lead them to eschew pretensions to beauty through ornament. They considered ornament and refinement to be justifications for, rather than antitheses of, material striving. Prizing material comfort and consumption, late nineteenth-century middle- and upper-class Chicagoans rationalized materialism by equating taste with culture. Not surprisingly, the buildings they produced did a great deal more than express their structural basis. Skyscrapers, then, are best understood as part of a larger cultural reformulation of urban commerce; furthermore, it is vital to comprehend the skyscraper as a downtown commercial space and as a workplace.

Chicago's nineteenth-century skyscraper boom occurred between 1880 and 1895. During these years the city's population increased at an annual rate of over fifty thousand persons. Growing numbers of business firms seeking growing amounts of office space in Chicago's

restricted downtown promoted the rise of the skyscraper. Between 1870 and 1890, the number of Chicago
lawyers, for example, rose from 629 to 4,241; in 1900,
7,032 errand and office clerks and 9,975 stenographers
and typists worked in Chicago (fig. 76), many of them in
skyscraper office buildings.[13] Skyscrapers frequently accommodated three to four thousand such workers, and
business visitors swelled the daily population of many
buildings two- or threefold (fig. 77). The expansion of
financial, legal, and administrative activity accompanied
Chicago's emergence as a regional and national trade and
industrial center. The skyscraper served an increasingly
complex economy.[14]

These workplaces, in turn, occupied a sharply constricted area of the city. Both natural and artificial boundaries hemmed in the office district. Lake Michigan
bounded downtown on the east, and as the *Chicago
Tribune* noted in 1888, the Chicago River was "considered a perpetual barrier against a westerly or northerly
expansion" of downtown. Bridges crossed the river, but
extensive lumberyards and other enterprises serving the
river port and dependent on water transportation lined
the riverbanks. Given the value of proximity for face-to-
face business, downtown was unlikely to straddle the
wide swathe of river and land committed to these inharmonious activities (plate 8). "Chicago has thus far had
but three directions, north, south, and west," wrote the
Tribune, "but there are indications now that a fourth is to
be added . . . zenithward. Since water hems in the business centre on three sides and a nexus of railroads on the
remaining, the south, Chicago must grow upward." Chicago could double its population and trade without
downtown "stirring from its tracks."[15]

When higher buildings became technically possible
and socially acceptable, land prices escalated to reflect
the possibility of higher density. "Skyscraper" buildings
burst the four- to six-story facade line of Chicago's 1870s
postfire commercial buildings, doubling and tripling
building heights to ten, twelve, and sixteen stories, low
by twentieth-century standards but unquestionably skyscrapers to contemporaries.[16] Thus, when Burnham and
Root's Masonic Temple building was completed in 1892,
it rose over three hundred feet above its corner at Randolph and State. On the entire rest of the block, one
building had an elevation of 96 feet and every other
structure was 76 feet high (fig. 78). On block after block
of downtown land, the jump in height from a stair-
climbing city to an elevator-riding city clearly impressed
local residents and visitors alike (fig. 79). To measure
these buildings by later skyscraper forms is to overlook
the crucial factor of relative height relations between

76. Stenographic Department, Sears, Roebuck and Company
(photo ca. 1904). Chicago Historical Society, ICHi 01638.

77. "How it might be if the 6,000 people in the Monadnock
should leave at one time." *Chicago Tribune* (24 February
1896), Sterling Library, Yale University.

78. Bird's-eye view of structures around Masonic Temple.
Building on the blocks bounded by Dearborn, Lake, Wabash,
and Randolph streets (1898 panorama). Detail of Poole
Brothers' Chicago Business District panorama. Library of
Congress.

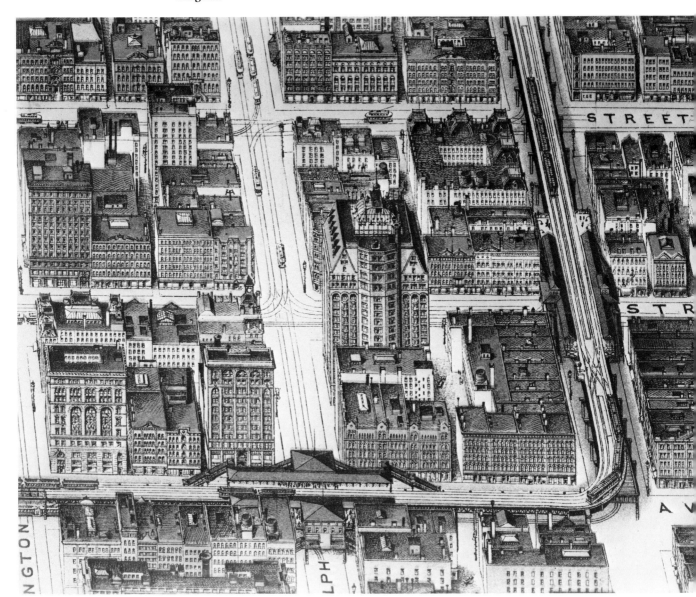

79. State Street looking north, Columbus Memorial Building in
foreground, Masonic Temple Building in the distance (photo
ca. 1895). Library of Congress.

individual buildings and their urban context. The low-rise, cast-iron and stone facades that had lined the commercial streetscape for decades (fig. 80) gave way to highrises, generally finished in brick, that visually dominated downtown.[17]

The concentration of highrises on the clearly ordered downtown grid gave the skyscraper, in Chicago, a prominence uncommon in other American cities. Chicago's "curiously situated" business district, "practically but nine blocks square," and a transit system that discharged workers at "practically a common center" contributed to the clustering of the city's skyscrapers. "The customs of the city have crystallized the tendency toward centralization to an unusual degree," wrote one observer.[18] In 1891 *Morris' Dictionary of Chicago* asserted, "Chicago is a city of 'skyscrapers.' While other towns may boast of isolated specimens of grand architectural creation, no other city can claim the general high average of elegant and massive buildings."[19] In New York, skyscrapers occupied isolated sites in a tangled street plan of the downtown area; viewing a cluster of New York skyscrapers often required a distant survey of the skyline, whereas a comparable view in Chicago required simply glancing along single streets (fig. 81). Chicago was more concentrated.

The fire that destroyed most of downtown Chicago in 1871 cleared the way for the city's distinctive and concentrated commercial district. Commercial developers in the postfire era generally did not have to contend with preexisting patterns of marginal residential, industrial, and commercial activity. The fire lent itself to the desire of city builders to order the cityscape. Early blocks of white-collar offices and shops had shared the bounded downtown area with one of the city's more squalid working-class neighborhoods. In 1872, real estate investor Malhon D. Ogden described the dramatic change that followed the fire: "That part of the burnt district north of Van Buren street, and between La Salle street and the South Branch of the River, before the fire, was covered with countless old rookeries and miserable shanties, occupied, for the past twenty years, as dens of infamy and low gambling dives, the resort and rendezvous of thieves, burglars, robbers, and murderers of all grades and colors, to the exclusion of all decency, or business purposes." The neighborhood had proved difficult to change because land owners allegedly received high rents from the "cutthroat class" for the "pestilent neighborhood [which] vouchsafed them from detection and arrest for crimes committed." The fire cleared the ground; it "destroyed the rookeries, drove the thieves

80. Cast-iron front commercial buildings on Lake Street in the 1870s (1876 photo). Chicago Historical Society, ICHi 22317.

81. Randolph Street looking east from La Salle Street. With the City Hall and Courthouse at right, this section of Randolph Street included three prominent skyscrapers: in the foreground, Ashland Block, Burnham and Root, 1891–92; middleground, the Schiller Building, Adler and Sullivan, 1891–92; and in the background, the Masonic Temple Building, Burnham and Root, 1891–92. Library of Congress.

away," and, with a not fully consonant conclusion, "large numbers of our heaviest wholesale and retail houses have located west of La Salle Street and north of Adams street."[20] The fire made way for a refined world of business, increasingly housed in skyscrapers and spatially and conceptually dissociated from other, less acceptable ways of making a living. The result was not merely a rebuilt, but a transformed, downtown.

Indeed, the creation of office buildings in downtown Chicago was linked to the creation of new forms of work established at a social and cultural distance from productive labor. The McCormick reaper works provides a good example of the evolving office activity that filled and shaped downtown skyscrapers. During the 1850s, McCormick's only two office clerks worked part time in the factory to fill their ten-hour days. In 1863, with business expanding, McCormick devoted special attention to office space as distinct from the rest of the plant. William S. McCormick wrote brother Cyrus that the office was "little and dusty—dirty I may say—unpleasant and unhealthy for so many men & boys as are generally in it. Agents come in and its full. Vault altogether insufficient. . . . Now we face the foundry & some days dust & smoke come into our windows. . . . We can not do our work well in present room & have no private room." To improve matters, William proposed building at the plant "a fine, large, airy & healthy office room" that was light, but shaded from the morning and evening sun, and exposed to the south breeze for ventilation.[21]

When the fire destroyed McCormick's office and works, the company acquired a new tract west of downtown and rebuilt its plant there. Uncertain about the practicability of establishing a general office at such an "inconvenient distance from the business center of the city," the McCormicks asked their St. Louis agent to survey companies similarly located and to report "the facts both in favor & against [an] office in immediate connection with the works."[22] The office remained at the works for the next six years. The unimproved character of the new plant location led one clerk to observe, "our office is fine and pleasant enough, but oh the getting here & getting home is terrible."[23] The McCormicks ran an omnibus to carry office clerks to the plant from the downtown transit lines.

By the late 1870s, the McCormick company had grown to the extent that it employed sixteen full-time clerks just to handle correspondence and accounts.[24] "The work is more and more complicated every year," wrote one manager in 1879, and office workers maintained "4 sets of regular books now besides all the auxiliary books." With the expansion of office work and office personnel came a distinct office culture within the company. The manager complained that McCormick's unwillingness to grant him more say in setting the clerks' salaries belittled him "in the opinion of all the office Men."[25] In 1879, a partnership dispute between Leander and Cyrus McCormick created a sharp split between Leander's supervision of manufacturing and Cyrus's control of legal, financial, and office affairs. Part of their settlement called for Hall, Leander's son, to remain "in the *manufacturing* dept. and not make trouble in the Office."[26] A growing distinctness of office and plant operations and the McCormick brothers' contest for control led the company to join other Chicago manufacturers who, forced by competition for centrally located land and rising prices, moved their offices downtown and their sprawling manufacturing plants to outlying areas.[27] The office corps moved downtown into the McCormick block on Dearborn Street (see fig. 90) and kept in contact with the factory using telegraphs, telephones, and express wagons. Similarly, Hannah Lay and Company long occupied an office at 78 Lumber Street, in the midst of the Chicago River lumberyards. With completion of the Chamber of Commerce, the company joined more than twenty other lumber dealers who opened offices in the new and lavish structure.[28]

As they joined financial, legal, and administrative businesses downtown, manufacturers devised images of their enterprises distinct from factory, foundry, and warehouse. In the 1880s, during a period of abrasive labor strife at the McCormick company and across the nation, McCormick changed its stationery letterhead. In the place of a bird's-eye-view etching of its manufacturing plant, complete with billowing smokestacks, the company substituted a bucolic image of a horse-drawn McCormick reaper. As police and striking workers battled in Chicago streets, McCormick's suppression of the factory suggested its discomfort in the face of the powers of production. New letterhead was a small token of a larger tendency in which businesses downplayed productive labor and the people who performed it.

The role of skyscrapers in the reformulation of business imagery is clear from contemporary guidebook descriptions. Earlier in the century, tourists routinely visited the extensive stockyards, lumberyards, and factories that made the city famous. Tourist guidebooks published in the late nineteenth century reflected and encouraged a realignment of the public's image of Chicago commerce. Skyscrapers received extended review in the tour itinerary, while factories receded into the background, into the guides' "general information" sections. One of the most thorough Chicago guidebooks of the 1890s, John J. Flinn's *Standard Guide of Chicago,* for example, passed over the major north-side manufacturing and warehouse district, commenting, "I will not ask you to penetrate this section now, but you can do so at your leisure."[29] In rather sharp contrast, Flinn recommended that city visitors devote at least a half day to the "study of" the Masonic Temple building, which at 22 stories and 302 feet was the world's tallest building (fig. 82). From the Masonic Temple's roof garden and observatory Flinn encouraged tourists to survey the stockyards, the Grant locomotive works, and the city's steel works, distilleries, breweries, and grain elevators.[30] This view of industry was distant, detached, and sanitized. The skyscraper, with its "comprehensive view," obviated much of the need to "penetrate" the city's industrial districts. In the context of many guidebooks, at least, the skyscraper took over as the representative expression of Chicago commerce.

Skyscraper designers accommodated their buildings to this role, shaping them as new workplaces and new symbols of a refined form of work. The specialized office building permitted architects, designers, and investors to devote money and imagination to the distinct needs and desires of the office tenant. Offices no longer simply occupied the left-over space above the storefront, in the corner of warehouses, or in the factory buildings. For discriminating tenants, the new office buildings combined office space with service, convenience, comfort, and even luxury. Building advertisements, rental agents' rhetoric, and architectural criticism all suggested that good commerce called for skyscraper designs with applied art and beauty.[31] If they omitted some of the intricate facade designs found in smaller, earlier, masonry and cast-iron-fronted commercial buildings, many architects who designed brick skyscrapers sought a balanced exterior massiveness, at times quiet and at times picturesque. Architect John Root felt that a profusion of exterior ornament was a "subtle means of architectural expression" that might go unheeded in "the midst of hurrying busy thousands of men."[32] In many buildings by Root and his contemporaries, starting with the entrance, energetic and artful interior design compensated for the occasional austere exterior.[33] Here, then, was not hostility to ornament but a judgment about where it would prove effective. Slighted by historians but notable to contemporaries, the entrances, lobbies, and elevators of Chicago skyscrapers were crucial to the ways that *clients*—building tenants and the business associates who visited them—experienced the buildings.

82. Masonic Temple Building, northeast corner of State and Randolph, Burnham and Root, architects, 1891–92 (1898 photo). Avery Library, Columbia University.

83. Honore Block, northwest corner of Dearborn and Monroe, Otis Wheelock, architect, 1870–71 (photo ca. 1870). Chicago Historical Society, ICHi 00993.

A host of examples establishes that architects thought artful embellishment was crucial to the clients' experience. Designs for skyscraper entrances differed substantially from those of earlier office blocks. The entrances to Chicago's preelevator office blocks were frequently unembellished. They occupied breaks between retail stores fronting the sidewalk. Some specialized office blocks built in the 1860s and 1870s did have more ornate entrances that were framed with porticoes and emphasized in the composition of the upper facade (fig. 83). The window bays immediately over these entries and the gables, finials, and cornice crowns at the roof ennobled both the entrances and the buildings. Truncated corners in such early office buildings as Van Osdel's McCormick and Reaper blocks (fig. 84), Jenney's Portland block, Burling and Whitehouse's First National Bank block, Frederick and Edward Baumann's Bryan and Metropolitan blocks, Willett's Times building, and the Union building (fig. 85) added grandeur to the entries while providing an unusual focal point for building designs constrained by the formal regularity of lots in the street grid. Yet many buildings lacked grand or even distinctive main entrances.

In contrast, the plan, form, and image of skyscrapers fostered a more emphatically monumental aesthetic for their entrances. More highly ornamented entries were encouraged by skyscraper plans: Earlier office blocks often supplemented front entrances with inconspicuous side entries, which provided stairs to the offices opening on the landings above the first floor. Multiple stairways and landings obviated the need for space-consuming systems of interior corridors. In skyscrapers, banks of elevators concentrated circulation within the building and thus favored a single entrance. Skyscraper form piled up hundred and thousands of tenants who generated traffic that required larger entrances. Finally, monumental entries seemed appropriate to the ennobled and perhaps ennobling symbols of commerce.

Many skyscraper architects adapted the technique of emphasizing entrances by articulating facade bays rising over them, even though the extraordinary height of many late nineteenth-century office buildings made this a less impressive compositional gesture than in lowrises. In the Chamber of Commerce building (see fig. 74), for example, the modest projection of the three windows above the Washington Street entrance portico was seen as an important part of the building's "claims to beauty."[34] Burnham and Root similarly composed many of their office buildings with an entrance pavilion established through a unified composition of the entrance with the bays and cornice above. In addition to adapting aspects of earlier office block design, many architects gave skyscraper entrances formal compositional autonomy. The monumental entrance on Solon S. Beman's nine-story Pullman building (1882–84) exemplified this approach (fig. 86). The main Adams Street entrance took the form of a triumphal arch, twenty-two feet wide at the sidewalk, elliptical in shape, and supported by massive columns with rectangular bases, polished red granite shafts, and foliated capitals. Connecting the two wings of the U-shaped building, the arch gave access to an eighty-foot-long court leading to an inside entrance and the building's elevators (fig. 87). Beman distinguished the Pullman's grand entry arch further by terminating it two and one-half stories above the ground, in marked disharmony with the building's dominant string courses.

Like the Pullman, Burnham and Root's Woman's Temple (1890–92) had an exterior light court and a massive,

84. McCormick's Reaper Block, northeast corner of Washington and Clark, John M. Van Osdel, architect, 1872 (1901 photo). Chicago Historical Society, ICHI 22308.

85. Union Building, southwest corner of La Salle and Washington, architect unknown, ca. 1868 (1868 photo). Chicago Historical Society, ICHI 01091.

86. Pullman Building, southwest corner of Michigan and
Adams, Solon S. Beman, architect, 1882–83 (photo ca. 1884–
89). Chicago Historical Society, ICHi 19460.

87. Pullman Building, Adams Street entrance court, Solon S.
Beman, architect, 1882–83 (photo ca. 1887–89). Chicago
Historical Society, ICHi 22309.

arched entrance (see plate 7). In Jenney's Home Insurance building (1883–85) the "grand entrance on La Salle Street [was] one of peerless beauty—a veritable marble hall, and a portal such as no palace in Europe can boast of."[35] (See fig. 88.) A massive portico supported by four Corinthian columns projected from the arched entrance. Hannah's Chamber of Commerce building had a similar entrance, a remnant of the 1872 building used by Chicago's Board of Trade. The architects, Baumann and Heuhl, had successfully appropriated the board's quasi-civic architecture for their building's owners and white-collar tenants. Buildings ranging from Burnham and Root's Phenix, Rookery (fig. 89), and Masonic Temple buildings to Sullivan's Stock Exchange to Boyington's Columbus Memorial to Holabird and Roche's Old Colony provided the grand entrance, an architectural gesture characteristic of leading civic and cultural buildings.

The palatial images established by grand entrances were heightened by skyscraper lobbies. Here tenants and visitors waited for the elevators in highly embellished spaces. The lobby of the sixteen-story Unity building, designed by Clinton J. Warren in 1891 (figs. 90–92), seemed like a "Fairyland" to one city guide:

> Entering through the great arch of the portal, rising to the height of a story and a half, the walls of the outer vestibule are composed of Numidian, Alps, Green and Sienna marbles. Over the inner door is an artistic screen of glass and bronze. Passing through the rotunda the eye is dazzled by its surprisingly brilliant beauty, designed in the style of the Italian renaissance. From the floor of the marble mosaic whose graceful design and harmonious color combinations are taken from the best example of the renaissance in the Old World, rises the first story by a marble balcony with marble balusters and balustrades.[36]

Corinthian columns with finely carved capitals, gold-leaf and silver chandeliers, and the silver-plated latticework of the elevators added to the grandeur. George H. Nesbot and Company of Chicago designed and fitted the Unity building lobby. Nesbot, A. H. Andrews, Henry Dibblee, the Winslow Brothers, and other designers and companies created a series of impressive lobbies in Chicago.

After passing through the columned portico of the Chamber of Commerce, visitors and tenants entered a vestibule with a mosaic floor and ceiling. Each of the thirteen floors had a different mosaic pattern. By the time they had finished the building, Burke and Company of Chicago had laid thirty-five thousand square feet of mosaic. The marble wainscoting quarried from the Italian

88. Home Insurance Building, northeast corner of La Salle and Adams, William Le Baron Jenney, architect, 1883–85 (1887 photo). Chicago Historical Society, ICHi 22312.

89. Rookery Building, southeast corner of La Salle and Adams,
Burnham and Root, architects, 1885–88 (photo ca. 1910). The
building to the left is the Home Insurance Building; at right is
the corner of the Illinois Trust and Savings Bank. Chicago
Historical Society, ICHi 19186.

90. Unity Building, east side of Dearborn between Randolph
and Washington, Clinton J. Warren, architect, 1891–92 (photo
ca. 1900). Lower building at left is the McCormick Block,
southeast corner of Dearborn and Randolph, John Van Osdel,
architect, 1872–73. Chicago Historical Society, ICHi 19117.

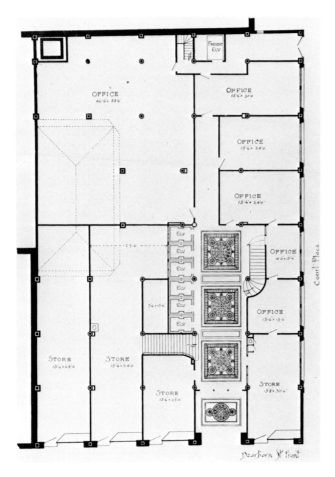

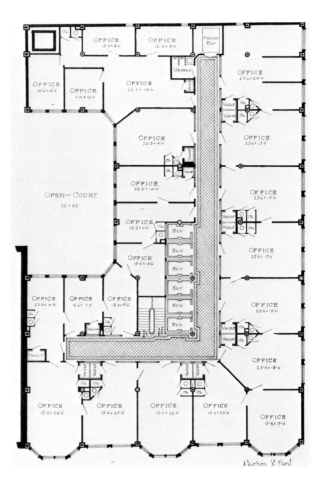

91. First floor plan, Unity Building, Clinton J. Warren, architect, 1891–92 (1892 photo). Avery Library, Columbia University.

92. Typical floor plan, Unity Building, Clinton J. Warren, architect, 1891–92 (1892 photo). Avery Library, Columbia University.

Appenines and finished in Belgium was put in place so that the grain would run continuously from one slab to the next.[37]

For all the impressiveness of its marble and mosaic work, it was ornamental iron that made the Chamber of Commerce building's interior stand out most notably. Intricately detailed cantilevered balconies ran around each of the building's thirteen floors, overlooking the enclosed light court (fig. 93). Winslow Brothers manufactured the balcony guards and rails using a lively foliated design that was adapted with some variations for elevator screens, columns, fascias, and stairways. Sun flooded in from the rooftop skylight and highlighted the bronze work (fig. 94). Winslow Brothers' interior metalwork, with complex patterns drawn from natural foliage and geometric shapes, embellished the lobbies of numerous Chicago skyscrapers, including the Auditorium (fig. 95), Caxton, Columbus Memorial, Home Insurance, Manhattan, Monadnock, Old Colony, Phenix, Pontiac, Rand McNally, Rookery, Stock Exchange, Tacoma, and Woman's Temple buildings.[38] Much like appropriated nature in city parks, the natural forms of foliated building ornament could evoke a consciousness of matters beyond the commerce of the urban market.

Essential to the skyscraper were elevators, first encountered by visitors and tenants as part of an embellished lobby. Nineteenth-century Americans commonly saw utilitarian machinery and technology as fit objects for aesthetic appreciation and hence for ornamentation.[39] Designers of skyscrapers gave elevators extensive artistic treatment. Winslow Brothers and other companies provided elevator enclosures and guards that flowed with lively floral and geometrical patterns in bronze and wrought iron (fig. 96). Chicago's Hale Elevator Company offered more than one hundred designs for its elevator cars, ranging in price from $200 to $2,000. Despite this variety, the manager of the Reaper office building, which had just installed Hale elevators, advised Mrs. Cyrus McCormick to engage her New York furniture designer to design the car for the elevator in her home. "It is so prominent an article that most Everybody sees and so much a matter of taste that I should think you need to give considerable time and attention to it yourself."[40] Although they enjoyed less choice than Mrs. McCormick, white-collar workers in the Reaper building and elsewhere might enjoy beautiful elevators. For manufacturers such as the Hale Elevator Company and their clients, elevators were not merely an important and convenient technology, they presented an opportunity to display taste and were ornamented accordingly.

Such ornament was simply essential to builders interested in winning tenants for offices, commanding high rents, and advertising themselves through skyscrapers. A skyscraper could vouch for the substance and good character of the company associated with the building either by name or by tenantry. In skyscraper lobbies, the Roman and Florentine mosaics and Halian, St. Sylvester, Italian, and other marbles covering floors, ceilings, walls, and staircases all alluded to such substance.[41] Contemporary descriptions of skyscrapers make clear the utility of such materials and workmanship as character reference; take, for example, this view of the Chamber of Commerce building: "From foundation to roof every inch of the building bears the impress of superb workmanship. There is not a trace of shoddyism about the structure. There is no veneering. There is no paint. Everything from the mosaic ceiling of the first floor to the Italian marble wainscoting of the thirteenth is real—not an imitation. No cheap substitutes have found their way into this work."[42] No doubt many business concerns were happy to associate themselves with such a building.

Businessmen were well aware of the potential of these buildings for advertising their wealth, taste, and even their public spirit. Two Chicagoans who brought motives beyond Mammon to the decision to build were Judge Van H. Higgins and his law partner, Henry J. Furber, owners of the sixteen-story Columbus Memorial building. Designed in 1891 by William W. Boyington, the building stood as one of the most notable skyscrapers in Chicago (figs. 97–99). Higgins and Furber sought to capitalize on civic celebration for commercial ends by constructing a skyscraper memorial to Columbus during the Columbian Exposition. Above the entrance arch on their building they placed a ten-foot-high statue of Christopher Columbus that had been sculpted in Rome. Other elements alluded to cultural associations, too. The grand entrance for the building was twelve feet wide and rose to the third story. The lower two floors of the building were clad in statuary bronze. In the mid-1860s Boyington incorporated a similarly impressive entrance arch in his design for Crosby's Opera House in Chicago (see fig. 50). Crosby's had provided a stage for grand opera and drama; its form bespoke its elegance and its enormous cost. In the Columbus Memorial, Boyington transferred architectural motifs from his civic and cultural building designs to a new monument of business.

Inside, the Columbus building was equally impressive. The halls and lobby were "richly decorated" in marble, mosaic and bronze, "treated in historical design"

93. Ornamental iron balcony and stairs, marble wainscotting, and interior light court, Chamber of Commerce Building, Baumann and Huehl, architects, 1888–90 (1894 photo). Avery Library, Columbia University.

94. Interior light court, Chamber of Commerce Building, Baumann and Huehl, architects, 1888–90 (1894 photo). Avery Library, Columbia University.

95. Marble and ornamental iron stairway, Auditorium Building,
Adler and Sullivan, architects, 1886–89 (1890 photo). Avery
Library, Columbia University.

96. Elevator enclosure screen, Woman's Temple Building, Burnham and Root, architects, 1890–92 (1894 photo). This ornamental iron elevator screen was manufactured by Winslow Brothers. Avery Library, Columbia University.

97. Columbus Memorial Building, southeast corner of Washington and State, W. W. Boyington, architect, 1891 (1895 photo). The memorial statue of Columbus is visible over the grand entryway on State Street. Chicago Historical Society, ICHi 22331.

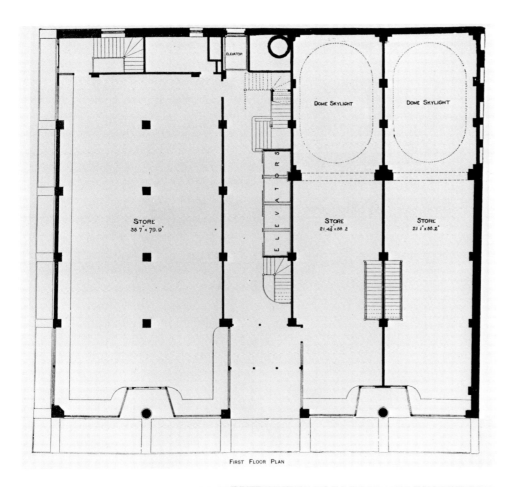

FIRST FLOOR PLAN

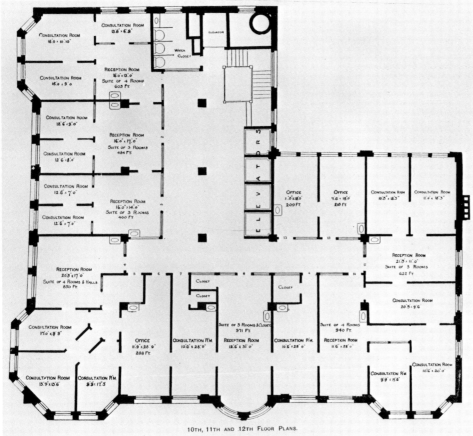

10TH, 11TH AND 12TH FLOOR PLANS.

98. First floor plan, Columbus Memorial Building, W. W. Boyington, architect, 1891. Avery Library, Columbia University.

99. Plan of floors ten through twelve, Columbus Memorial Building, W. W. Boyington, architect, 1891. Avery Library, Columbia University.

(figs. 100, 101). Its owners proudly declared, "These hallways will be finished in a style unequalled by any hallways in any business building in Chicago or elsewhere." Boyington adopted a range of architectural motifs from the Spanish Renaissance. An octagonal corner tower on the roof supported an illuminated glass globe of the world on which a cut jewel indicated the location of Chicago. Mosaics commissioned from the Venice-Murano Company of Italy filled the back walls of the two skylighted ground-story stores facing onto State Street. One mosaic depicted the "Landing of Columbus," the other showed "Columbus before Ferdinand and Isabella"[43] (fig. 102). Here the appropriation of civic form by commercial interests went beyond architecture to civic ritual. The skyscraper stood as a celebration of both the discovery of America and of Chicago's triumphant commemoration of the event in the Columbian Exposition.

At the root of the Columbus Memorial project lay a determination on the part of the owners to go beyond the necessities of real estate operation to advertise their own wealth, solidity, and public spirit. As the *Chicago Tribune* reported, "A million dollars in a large sum to expend on a lot 100 × 90 feet, when a building costing from $650,000 to 750,000 would in all probability bring in the same rental."[44] Status in the Chicago community was based firmly on wealth but also on the tasteful expenditure of it. According to a contemporary biography, Van H. Higgins was a "typical western man." Born in 1821, Higgins was the son of a Genesee County farmer. He first traveled to Chicago in 1837, settled there permanently in 1852, and proceeded to make a fortune at the bar and in various manufacturing and insurance enterprises. In 1891, at age 70, he initiated the Columbus Memorial. Higgins's biographer credited him with prescience about Chicago's ability to challenge eastern cities as a center of civility and refinement. Higgins had "seen with a vision clearer than most men, not only the probabilities but the possibilities of this Great West, and what a quarter of a century and more ago he so clearly saw, and what he so confidently prophesied, he has diligently worked to realize."[45] Higgins's partner, Henry J. Furber, was younger. Born in 1840, he moved to Chicago in 1879. A man of culture, Furber had "princely" tastes and owned "collections of art and literature of the rarest and most expensive character in the world."[46] Furber explained that the Columbus Memorial design heeded the advice of people who insisted "that we do something for the city. They said that instead of building a structure after the drygoods-box style followed by most Chicago

buildings, we ought to erect a structure which would be an ornament to the city. In other words they suggested that in the improvement of the corner we blend some sentiment with cold business policy."[47]

Such criticism of "dry goods style" was not generally aimed at skyscrapers. The grand entrances, embellished lobbies, rich marbles, mosaics, hardwoods and ornamental ironwork were all quite common among these tall buildings and set them off from earlier and plainer structures in the downtown area. Admiring the "elegant finish" of Burnham and Root's Phenix building, "a building devoted to purely commercial purposes," one critic predicted that its example of beauty would be followed "until art would find a place where before rough walls and plain finish was considered all that was necessary."[48] Burnham and Root's Masonic Temple building, according to one critic, incorporated "all the arts of the present century . . . to embellish its interior and give it an attractive exterior."[49] Beyond the grand entrance stood a whole host of modern planning, technical, and service innovations, *arts* that distinguished the skyscraper from earlier office blocks. To some, skyscrapers heralded a new and refined style.

In turn, architectural refinement reflected upon the work taking place in skyscraper offices and on the workers who performed it. Chicago businesses faced the world less at the factory and more in the officeplace, presenting themselves in sales and advertising through a middle-class, white-collar workforce that was more presentable than the working class. For professionals, managers, and lowly clerks, removal from industrial work entailed a social dignity that was enhanced by the offices themselves (figs. 103, 104).[50] Beyond the elevators, these office tenants found a variety of modern technological systems combining artistic appearance and mechanical precision—gas and electric light fixtures, steam radiators, air vents, bathroom fixtures, and door hardware.

The skyscraper's supporting technologies and architectural embellishments promoted the light, ventilation, cleanliness, comfort, healthfulness, and beauty of office accommodations. In the early 1890s, new skyscrapers such as the Monadnock, Unity, Woman's Temple, Masonic Temple, and Ashland filled their offices with tenants anxious to move out of "old, second, third, and fourth class office buildings."[51] Tenants indifferently cast aside old workplaces, and like urban flat dwellers took leave of a place on May 1st moving days with only the "slightest encouragement" and few regrets.[52] In 1886, for example, the publishers of *Railway Age* moved their offices into the recently completed Home Insurance building.

100. Lobby, elevators, Cutler mail box, Columbus Memorial Building, W. W. Boyington, architect, 1891 (1894 photo). Avery Library, Columbia University.

101. Main stair and elevator enclosures, Columbus Memorial Building, W. W. Boyington, architect, 1891 (1894 photo). Winslow Brothers provided the ornamental ironwork for the building; note the use of mosaic in the upper halls as well. Avery Library, Columbia University.

102. Skylight in first floor store, Columbus Memorial Building,
W. W. Boyington, architect, 1891 (1894 photo). The main light
court of this building rose above this skylight, manufactured by
Winslow Brothers. The Venice-Murano Company of Italy de-
signed the mosaic scene of "Columbus before Ferdinand and
Isabella." Avery Library, Columbia University.

103. Board of Trade, Jackson at foot of La Salle, W. W. Boyington, architect, 1881–88 (1893 photo). In the district around the Board of Trade stood, at left, the Home Insurance Company Building and Rookery Building; at right center, the Insurance Exchange Building, Burnham and Root, 1884–85. Chicago Historical Society, ICHi 00255.

104. Main trading room at the Board of Trade (1905 photo). Library of Congress.

They had considered their old building first class when they occupied it a few years earlier. The six-story office block at 103 Adams Street had a modern elevator and a small light well for light and ventilation, but it suffered "greatly by comparison" with the new skyscrapers. Although the structural system supporting the Home Insurance building has stirred great debate among historians, it made no impression at all on the magazine's publishers.[53] They were impressed with the marble used for floors, ceilings, stairs, and wainscoting, with the hardwood office interiors, the rapid elevators, and the provision of natural light, gas and electrical illumination, steam heating, and fire and burglar protection. These fine features and the "high character of its tenantry" made the Home Insurance the "finest and most complete office building in America . . . supplied with every convenience that modern skill and lavish expenditure of money could command . . . a model of beauty, convenience, and comfort."[54]

In spite of some highly decorative exteriors, the interiors of the office blocks of the 1860s and 1870s appeared crude by skyscraper standards.[55] In the "first-class" office blocks of the 1860s, such as the Larmon and McCormick blocks, physicians, lawyers, real estate agents, jewelers, and bankers occupied offices located above assorted storefront businesses. The offices depended on fireplaces and stoves for heat, gas jets for light, and open windows for ventilation. Dark corridors, long flights of stairs and, eventually, rather slow elevators provided access to offices. City water was available but in the 1860s, the Larmon and McCormick blocks continued to rely on privy toilets, with lime added for sanitation. Tenants cleaned their own offices. Starting in 1867, the Pinkerton detective service included the Larmon and McCormick blocks on its nightly rounds, but no special watchman or guard and only one part-time janitor worked inside the buildings.[56]

By contrast, skyscrapers of the 1880s and 1890s furnished tenants with a wide variety of modern technologies well in advance of their availability to other city buildings and neighborhoods. In the early 1880s, electric lights replaced gas illumination in office buildings, long before the transition took place in other parts of the city. Skyscrapers centralized systems of steam heating and mechanical ventilation, removing from the tenant the responsibility for heating and ventilating offices. Extensive water and sewerage systems won suitable architectural expression in large, elegant toilet rooms (fig. 105) and eventually furnished water and toilets on every office floor. Hot and cold running water became a standard

feature of many office suites in the 1880s. A succession of electric bell signal systems, pneumatic message tubes, and office building telephone exchanges helped tenants to communicate with other parts of the building and the city. While urban residents struggled with the vagaries of intraurban transportation, elevator technology moved ever greater numbers of people more quickly through increasingly tall buildings. Letter chutes moved mail from nearby each office to a central collection box almost effortlessly. Throughout the skyscraper boom, the fittings of a *modern* and *complete* building changed as higher levels of artistic finish and mechanical services were offered in the competition for tenants.

In the design and promise of the skyscraper, nothing proved more fundamental than the provision of higher quality natural light for offices. Investors, architects, tenants, and architectural critics carefully scrutinized these provisions.[57] Architects expanded office block light shafts into tremendous skyscraper light courts. Skyscraper architects and investors sacrificed large amounts of interior space to provide for natural light, and the desire for greater light helped determine not only the floor plan of the skyscraper but also the building site selected.

Some of the lower office blocks constructed in the 1870s also evidenced concern over improved natural light and ventilation. In 1872, Boyington's five-story Superior block, constructed on Clark Street opposite the courthouse, incorporated a novel plan for increasing natural light in offices. The building lot was relatively deep, 67 feet by 161 feet, and the building shared party walls with adjacent buildings. The standard solution for lighting the back offices would have been to construct interior windows opening into light shafts. Instead, Boyington broke the project into two sections separated by a glass-enclosed light court, approximately 30 feet by 40 feet, that provided superior natural light to back offices. Boyington located the only elevator in the front section of the building; bridges connecting the front and back sections spanned the courtyard at all levels. Sacrificing some rental space for better light made the Superior block the "best appointed, and most elegant business block" in Chicago, and brokers, insurance agents, and lawyers eagerly moved into the building.[58] The Farwell block, built in 1870 after designs by George H. Edbrooke, also incorporated an impressive light court. The building was constructed on Arcade Court behind John V. Farwell's Republic Insurance block. It received light from both Arcade Court and from an alley. A substantial central court, however, lighted the halls on the building's six floors; glass partitions admitted light from these halls to

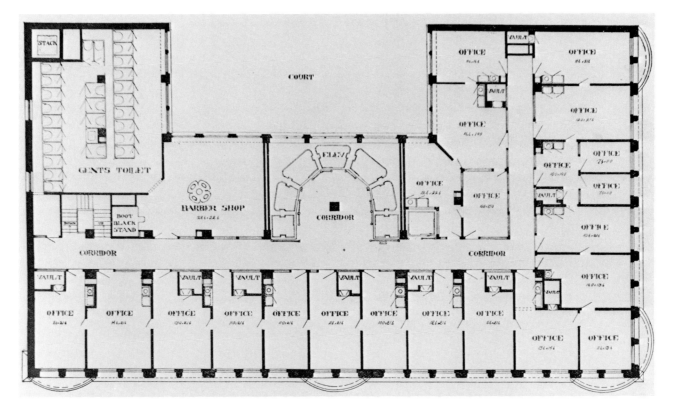

the part of each office farthest from outside windows.[59]

When office building construction resumed after the 1870s economic depression, the concern for improved natural light continued to influence office design. Architects expanded light shafts farther and in some important designs, they moved standard modest light courts from the back of an office block into the center of the building. The modern light court became an integral and impressive part of the building's architecture, enjoyed by tenants and visitors alike. Two office buildings designed in 1881—Burnham and Root's Chicago, Burlington, and Quincy Railroad at Adams and Franklin, and Burling and Whitehouse's First National Bank at Monroe and Dearborn (fig. 106)—introduced grand office light courts in Chicago. In Burnham and Root's design, a light court of over 100 feet by 50 feet stood at the center of a six-story building that covered a plot of land only 176 feet by 122 feet. Iron galleries encircled the interior court on each floor, giving access to the offices from the building's stairs and elevators. Here light and air streamed into offices from windows on both the street and light court.

In the First National Bank building, the large central light court served a double purpose. It lighted the ground-story banking room (fig. 107) as well as the inside offices on the upper floors of the six-story building. The light court covered 113 feet by 19 feet in a structure that measured 190 feet by 95 feet. The officers of the First National Bank considered their building to be responsive to the "demand for perfect office quarters." The

105. Plan of fifteenth floor, Ashland Block, Burnham and Root, architects, 1891–92. Office buildings like the Ashland Block advertised their extensive and sumptuous toilet rooms and barber shops. Avery Library, Columbia University.

106. First National Bank Building, northwest corner of Dearborn and Monroe, Burling and Whitehouse, architects, 1881–82 (photo ca. 1895). At left, on Monroe Street, is the Montauk Building, designed by Burnham and Root, 1881–82. Chicago Historical Society, ICHi 22310.

107. Banking room of the First National Bank Building, Burling and Whitehouse, architects, 1881–82 (photo ca. 1885). Chicago Historical Society, ICHi 22299.

building's grand entrance with its Doric portico and two polished columns, the lobby with its columns of New Brunswick red granite, its Tennessee marble wainscoting, and its imported tilework installed by Henry Dibblee and Company, the Hale elevators, the ornamental bronze work—all contributed to this ideal. The flood of natural light mattered, too.[60]

Chicago's Knisely and Company built the skylights for both the Chicago-Burlington and the First National buildings. Knisely worked with Hayes patented metallic ventilating skylights, which—according to advertising and newspaper accounts—made the skylighted office building practical. Worries about leaking skylights had deterred earlier builders from their use and contributed to the problem of "Close, Dark and Dingy buildings." With Hayes's patented new methods of setting glass and draining skylights, said promoters, "the modern building is sure to be Light, Airy and cheerful-looking, and always calculated to secure the best tenants."[61] In 1884, the Open Board of Trade building, designed by architects Wheelock and Clay, also included a Hayes patent skylight 65 feet by 85 feet in a building that covered only 100 feet by 105 feet. As in the First National Bank, one skylight covered a large, open, ground-floor trading room, and a second skylight covered the roof. As in the Chicago, Burlington, and Quincy Railroad building, access to the offices on the five upper floors was possible along iron galleries overlooking the open light court.[62] These galleries and open courts dramatized the entire matter of natural light, which served as the cornerstone of attempts to "secure the best tenants." In Burnham and Root's Rookery building, the ground story lobby and shops were lighted by an impressive skylight (fig. 108), and glazed bricks lining the interior walls helped reflect natural light into offices on the upper floors.

An 1888 *Tribune* article, "The Sacrifice of Space for Light," reported: "Dark rooms will not rent, and it therefore does not pay to construct them. The question is how to get the greatest amount of rentable space with the smallest cubic contents, and each lot presents its own peculiar problem. The old practice was to cover the entire lot, and the consequence was dark rooms in a considerable proportion of the space." Dark offices rented only at very low rates, and "buildings constructed according to the latest ideas have readily taken tenants away." Thus, experts criticized the narrow light shaft initially planned for the Chamber of Commerce building as insufficient. The *Tribune* sympathized with Hannah, Lay and Company which, having purchased the lot for $625,000, understandably balked at throwing large areas

into light courts. Yet the paper predicted that "careful figuring on rent is apt to lead to the conclusion to leave 'a little more' out of doors."[63]

In its final design, the Chamber of Commerce's 35 foot by 108 foot light court turned the ideals of light and ventilation into a grand architectural gesture. Tenants and visitors passed through the columned portico and the mosaic and marble vestibule, past the ornate elevator enclosure and into a two-hundred-foot-high light court. Here "a perfect flood of light penetrates the central court, so that the interior of the building is almost as brightly illuminated as the exterior during the day."[64] The design had powerfully captured and framed the sky, enhancing both the aesthetic and practical qualities of the architecture. Few buildings contained such a prospect. In 1892, Burnham and Root's Masonic Temple building opened with an interior light court that rose through ornate balconies and by office windows twenty-two stories—more than 302 feet—to a glass skylight on the roof (figs. 109–12). In these skyscrapers, interiors now competed with exteriors. The Chamber of Commerce appeared in relatively simple visual terms, "a tall box, all its grandeur being found within."[65] In creating that sense of grandeur, light played no small part: "with the owners light has been a prime consideration."[66]

Freestanding buildings occupying large plots of land completely surrounded by streets and alleys contained the city's grandest nineteenth-century interior light courts. The Rookery, Masonic Temple, Chicago Burlington and Quincy Railroad, First National Bank, and Chamber of Commerce all stood as islands. In these buildings, light courts supplemented the ventilation made available from outside windows. Other plots of land in Chicago's grid suggested a different solution. Narrow Dearborn Street blocks and lots south of Jackson Street proved popular because they provided ideal light without any light courts or light wells. Such buildings as the Monadnock (1889–92), designed by Burnham and Root; the Fisher (1897), designed by D. H. Burnham and Company; the Old Colony (1894), by Holabird and Roche; and the Manhattan (1891), by Jenney, occupied narrow blocks and provided offices flanking central corridors that were well lighted from outside.

The skyscraper's steel-frame structure facilitated natural light by permitting a larger window area than was possible with load-bearing walls. *Chicago* windows, composed of a fixed sash flanked on either side by smaller movable sashes, flooded offices with light while giving easy control over air circulation. The bay windows that so prominently protruded from skyscraper facades also contributed to the improved lighting and

108. Skylighted lobby, Rookery Building, Burnham and Root,
architects, 1885–88 (photo ca. 1895). Avery Library,
Columbia University.

109. Lobby at the base of the central light court, Masonic Temple Building, elevators at rear, Burnham and Root, architects, 1891–92 (1892 photo). Avery Library, Columbia University.

110. Main stair, elevator enclosures, balcony rails, Masonic Temple Building, Burnham and Root, architects, 1891–92 (1892 photo). Avery Library, Columbia University.

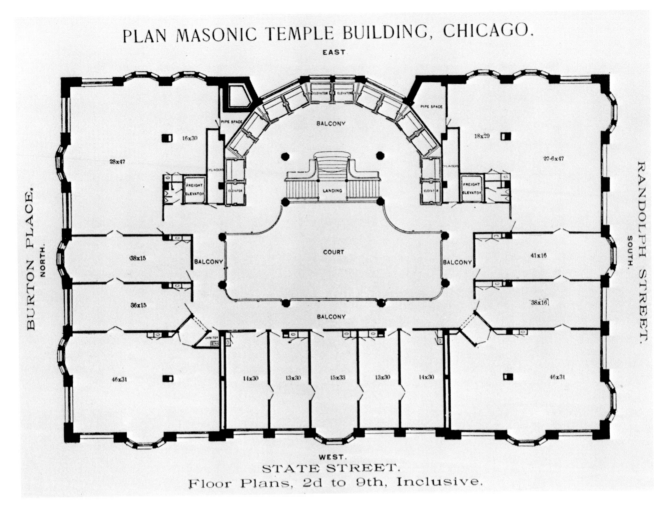

111. Plan of floors two through nine, Masonic Temple Building,
Burnham and Root, architects, 1891–92. These floors were
planned to accommodate retail shops with corridor circulation
around the balcony of the central light court. Avery Library,
Columbia University.

PLAN MASONIC TEMPLE BUILDING, CHICAGO.

Floor Plans, 10th to 16th, Inclusive.

112. Plan of floors ten through sixteen, Masonic Temple Building, Burnham and Root, architects, 1891–92. These floors were planned for offices. An inner tier of offices had windows facing the central light court; an outer tier of offices had windows opening onto adjacent streets. Avery Library, Columbia University.

ventilation provided to office interiors (fig. 113). Thus from skyscraper site to building layout to facade design, architects and builders carefully considered natural light.[67]

The centrality of light in the aesthetic that shaped Chicago skyscrapers reveals several points. It was an aesthetic that arose in part from considerations of utility: The widespread demand for natural light had to do with the needs of white-collar workers who wrote, typed, filed, and otherwise processed ever-increasing amounts of paper. Just as important were the implications of natural light and fresh air for the cultural definition of the office environment, the social construction of the work—which is to say, the matter of class. Light seemed crucial to white-collar work because that work was being defined as respectable, appropriate to the refined and cultivated classes of the city. Thus, the *Tribune* spoke of builders of the Chamber of Commerce: "Their work may not embody Sweetness enough to satisfy Matthew Arnold, yet it should be Light itself."[68] The *Tribune's* language testifies to contemporaries' acute consciousness of a cultured ideal and the degree to which that ideal informed skyscraper form and style. Office workers fancied themselves "people of culture."[69] It was a new class in the making—not some abstract entity such as architecture or construction—that needed light, air, clean hands, and clean collars.

Here, too, was an aesthetic that created a necessary connection between commerce and culture, denying their incompatibility and suggesting instead that refinement might emanate from tasteful workplaces. From this perspective, it is possible to appreciate the ornamented entryways, lobbies, and elevators as fully of a piece with other aspects of skyscraper buildings, not moments when designers lost sight of the centrality of function and structure, but rather moments when those designers fulfilled ideals that justified commerce with ornament. In no case was the appellation "an ornament to the city" a pejorative phrase.

Interestingly, the enthusiasm for improved natural light derived appeal from popular medical and cultural beliefs that also influenced the planning of other leading modern urban forms. Indeed, the concern of designers for natural light suggests somewhat unexpected connections between skyscrapers and more traditional plans for urban parks and suburban neighborhoods. Notwithstanding considerable progress in the new field of bacteriology, which linked microscopic organisms with specific diseases, older, miasmatic theories of disease persisted through the end of the century. Many people

113. Reliance Building, southwest corner of State and Washington, Burnham and Root/D. H. Burnham and Company, architects, 1890 and 1894–95 (photo ca. 1897–1904). Chicago Historical Society, ICHi 01066.

1. Chicago Boulevard and Park System, 1881 (1881). Library of Congress.

2. and 3. Promenaders in fashionable clothing, 1828–36. Fashion plates from National Museum of American History, Smithsonian Institution.

4. Grand Boulevard (now Martin Luther King, Jr., Drive) looking north from Fifty-first Street. Postcard (ca. 1910) from author's collection.

5. Jules Guerin, *Chicago: Proposed Plaza on Michigan Avenue, West of the Field Museum of Natural History in Grant Park, Looking East from the Corner of Jackson Boulevard,* published as plate 128 in *Plan of Chicago,* 1909. Reproduced from an original painting, Chicago Historical Society.

6. Louis Edward Hickmott's perspective view of the Court of Honor, World's Columbian Exposition, 1893. Chicago Historical Society.

7. Woman's Temple Building, southwest corner of La Salle and
Monroe, Burnham and Root, architects, 1890–92. Postcard
(ca. 1900) from author's collection.

8. Panoramic view of Chicago, Currier and Ives, 1892. Library
of Congress.

9. Jules Guerin, *Chicago: View Looking West over the City, Showing the Proposed Civic Center, the Grand Axis, Grant Park, and the Harbor,* published as plate 87 in *Plan of Chicago,* 1909. Reproduced from an original painting, courtesy of Patrick Shaw.

10. Jules Guerin, *Chicago Railway Station Scheme West of the River between Canal and Clinton Streets, Showing the Relation with the Civic Center,* published as plate 122 in *Plan of Chicago,* 1909. Avery Library, Columbia University.

believed that miasmas, polluted air and noxious odors arising from the decay of animal and vegetable material, caused human disease and explained epidemics.[70] Popular notions of health suggested that natural sunlight and freely circulating air were the surest cure for miasmas and the unhealthy "sweat of the city." Sun and air were the "great disinfectants which nature provided to make habitation" of Chicago's swampy site possible.[71]

Extending toward the sun, elevating tenants above earthly miasmas, making abundant provision for natural light and mechanical ventilation, skyscrapers appeared to many nineteenth-century Chicagoans to be allied with more natural landscapes such as those in parks and suburbs. People besides office tenants enjoyed the elements of skyscrapers that made them urban enclaves. Roof gardens on the Masonic Temple and the Great Northern Hotel opened in the 1890s and became resorts for many visitors seeking views of the city, refreshments—and during the summertime—concert music or vaudeville theater. One guide reported, "on the warmest evenings these open air amusement houses are delightfully cool, and swept by the lake breezes."[72] The middle- and upper-class tenants of the skyscraper achieved an alliance of these seemingly opposite landscapes—skyscrapers and parks, offices and suburbs, work and leisure. These same people who responded enthusiastically to well-lighted skyscraper offices proposed, built, and used urban parks, and embellished residential districts and suburban communities.

Finally, human services as well as technology and aesthetic design made the skyscraper attractive to white-collar office workers. Without leaving the building, tenants had access to a variety of stores, restaurants, and professional services such as barbers, cigar and newspaper stands, tailors, doctors, dentists, lawyers, and bankers. Owners of the Columbus Memorial and the Ashland buildings appealed to particular constituencies by providing specialized medical and legal libraries (fig. 114). Aside from the convenience of doing business with the building's other tenants, building managements offered a vast array of personal services. Doormen, elevator operators, messenger boys, and building detectives were the most visible representatives of these services. Janitorial staffs cleaned the public areas of skyscraper buildings and for an additional fee cleaned tenant offices. Engineers, electricians, and mechanics worked to keep the buildings' mechanical systems functioning properly. Building managers coordinated and oversaw the entire operation, while rental agents attracted and helped accommodate tenants. The use of nonstructural wall partitions in many new buildings permitted tenants to obtain office space that met their particular requirements. Human or mechanical failure, such as slow elevator service, cold offices, lighting failures, broken promises concerning prospective tenants or impertinence on the part of doormen or elevator operators, all provoked loud protests from tenants.[73]

Even those employees primarily committed to the successful operation of a skyscraper's technology could contribute to an atmosphere of efficient and personalized service for tenants. Earlier office blocks had generally employed only a rental agent, a janitor, and occasional outside help to perform needed repairs. In the Larmon and McCormick blocks, less than 5 percent of rental revenue was spent on service employment during the 1860s. By contrast, at the Board of Trade building, over 20 percent of office rental revenue went to pay building employees. In 1886, for example, the Board of Trade employed 37 men and 4 women—including 7 elevator operators, a chief engineer, electrician, steamfitter, machinist, blacksmith, carpenter, 3 firemen, a coal passer, a laborer, and 4 day janitors who washed windows, light globes, and lamps, wound the clock, and cleaned both ironwork and sidewalks. At night, 9 men and 3 women spent 12 hours cleaning 81 rooms, 9 water closets with 64 flush toilets, the halls and the corridors. The nighttime crew scrubbed water closets every day, the corridors twice a week, and—for an additional $7 to $14 monthly—offices once a week. Watchmen guarded the building twenty-four hours a day. Some skyscrapers even provided janitor quarters on the roof of the building. At the Board of Trade, a manager had general charge of all the employees and of rental, bond, and insurance accounts.[74] Middle-class tenants could be secure in their status; they were members of a class that was served.

Distinguishing even lowly clerks from blue collars, scholars have noted, was part of a rationalizing effort, a means of controlling the new workforce.[75] For example, contemporary reviewers of the Pullman building noted that it was "much more extensive and elaborate than required for the purposes of office accommodations. The building's "palatial" character reflected the "same spirit" evident in the model company town at Pullman.[76] In both venues, Pullman revealed a belief that clean and pleasant surroundings would promote worker productivity. In 1873, the company built an office building with a restaurant, library, and sitting rooms for employees and their families. By "cultivat[ing] society, as it were, among the employees" and by making company headquarters "attractive," Pullman hoped that the building would be "productive of harmony and good feelings, while it will

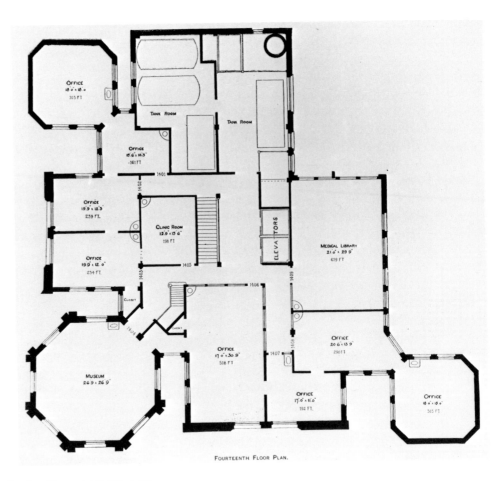

FOURTEENTH FLOOR PLAN.

114. Plan of fourteenth floor, Columbus Memorial Building, W. W. Boyington, architect, 1891. Higgins and Furber, the building owners, built a specialized medical library, museum, and examining rooms on the top floor of their building in an effort to attract medical specialists as tenants. Avery Library, Columbia University.

interest them more in the work for which they are employed . . . [and] make them more useful."[77] Perhaps an embellished office building with a grand, ennobling entrance might do more than indicate the presence of refined workers; it might actually cultivate them. Ever more difficult to achieve at the factory, harmony could be sought at the office.

Skyscrapers thus codified a more benign image of modern commerce, narrowing, focusing, and purifying images of work. They were refinements of the downtown area, all the more powerfully so given the general clearing of downtown blocks accomplished by developers both before and after the fire of 1871. Skyscrapers helped redefine downtown as an acceptable arena for both respectable gentlemen and ladies. They achieved monumental status, calling for special homage and awed respect transcending their workplace character. Skyscrapers became the object of special pilgrimages to ornate lobbies, sumptuous restaurants, and rooftop gardens.[78] As a center of administration and coordination, the office building contained little visible production. Flinn's guide observed, "What all the people who occupy the offices do will be a source of wonder to the visitor . . . but as they are all compelled to pay high rentals it is presumed that they are doing something to coax the almighty dollar in their direction."[79] Work itself could seem comfortably abstract within skyscrapers.

In fact, tall buildings stood among the late nineteenth century's most dramatic and expensive custom-built products, assembling parts from literally hundreds of subsidiary manufacturers. Like a plethora of new mass-produced consumer products, skyscrapers were promoted as a means to foster human efficiency, comfort, and pleasure and to buttress a rising standard of living. As a new organizing form for business life, as an eager patron of diverse mechanical innovations, as a competitive, custom-made, revenue-producing structure, the skyscraper compelled architects to consider judiciously their complex designs, combining elements of both beauty and utility.

Architects' awareness of the complex nature of tall office buildings appeared both in their skyscraper designs and in their writings about them. Emphasizing cultural and aesthetic simplicity over complexity, Condit and Giedion focused narrowly on a handful of interesting, somewhat austere, structurally expressive commercial buildings. By designating their architects as a *school* and giving these buildings status as a *style,* they obscured important aspects of skyscraper history and the history

of urban culture. In the nineteenth century, even architectural critics who lauded the relative plainness and simplicity in Chicago skyscrapers questioned the need for structural expressionism: "There is no greater folly than to maintain that Chicago architects have not found a suitable means of architecturally expressing modern steel construction," said one writer in the *Tribune.* "The steel skeletons of the office buildings need be no more expressed than the bones of the critics rash enough to father this statement."[80] Even if architects considered the expression of skeleton construction to be essential, distinct styles could have developed. For example, depending on the placement of spandrel and pier panels, a skyscraper's steel skeleton could express itself as vertical, horizontal, or in balance. These diverse solutions, all of which appeared in Chicago, could only questionably represent a single style or school. In fact, many Chicago architects and critics spoke as if a new style had emerged; yet simple massive scale defined it more closely than any narrowly conceived aesthetic treatment. In 1886, the *Tribune* reported that "when Chicago takes old Rome's arches and sticks on top of them a skyscraper block containing 5,000 rooms, a cafe, an opera-house, a barber-shop, and a billiard saloon, the whole thing is an architectural triumph and justly belongs to the new school of Chicagoesque."[81] A few years later another commentator declared, "The title, commercial style . . . may be said to embrace, generally, all modern houses over seven stories in height."[82]

As commerce provided a foundation for the development of Chicago architecture and culture, it was rarely conceptualized as an end in itself. The city building elite hoped that a refined, civilized life and art would emerge from modern commercial society and remained dissatisfied with forms that seemed to express only Mammon. Louis Sullivan probed this aspect of the skyscraper concisely when he wrote, "Problem: How shall we impart to this sterile pile, this crude, harsh brutal agglomeration, this stark, staring exclamation of external strife, the graciousness of those higher forms of sensibility and culture that rest on the lower and fiercer passions?"[83] Modern twentieth-century architects and their critical champions traced the lineage of their designs back to Sullivan, but they overlooked or condemned Sullivan's exuberant ornament and the transcendent aspirations of his proposals for commercial architecture. In writing on the design of the tall office building, "artistically considered," Sullivan echoed a familiar theme from cultural commentaries of the time; he equated commerce in its utilitarian aspects with a certain crudity. Art and beauty were intended to refine commerce's utility. Of the skyscraper, Sullivan wondered, "How shall we proclaim from the dizzy height this strange, weird, modern housetop the peaceful evangel of sentiment, of beauty, the cult of a higher life?"[84]

Part of the answer was the continuing and conscious association between cultivation and ornament: A skyscraper should be "every inch a proud and soaring thing," expressive of sentiment by means of embellishment. For example, the first floor and the entrance, where public attention centered, should be designed "in a more or less liberal, expansive, sumptuous way—a way based exactly on the practical necessities, but expressed with a sentiment of largeness and freedom."[85] In skyscrapers designed by Sullivan and his contemporaries, "sumptuous" ornament, found especially at the top and bottom of the building, and in grand entrances and embellished lobbies helped mediate the distance between commerce and culture, function and form. Structural expressionism informed only certain aspects of certain buildings because commerce and utility formed only a part of the urban culture they expressed.[86]

William Le Baron Jenney, the architect of the Home Insurance building and other notable Chicago skyscrapers, shared something of Sullivan's cultural outlook and idealism. Initially he had balked at the idea of settling in the unrefined Midwest. Seeking employment in the east, he wrote to Frederick Law Olmsted after the Civil War. "Art & Construction have been my serious study since '58, and it is my wish to continue it. In the West there [is] little knowledge and little desire for Art: besides one here must live within themselves without the means of profiting by the works of others."[87] When Jenney took up permanent residence in Chicago, he did not resign himself to an artless, commercialized world but rather promoted beauty and refinement in art and construction. Despite zones of structural expression in some of his buildings, Jenney did not think that the mere expression of technology or utility was sufficient. He adhered to "an old and well-established principle in architecture, to ornament construction, never to construct for the sake of ornament." Styling himself an arbiter of artistic taste, Jenney crusaded for art education, galleries, and intelligent art criticism that would in turn give art "a place in the thoughts of the people as a co-laborer and equal with her sister, Agriculture, Commerce, and Manufacture."[88]

Jenney's interest in the cultural associations of his design work as well as his own understanding of visual perception as it related to architecture influenced his ornamental treatment of the skyscraper. As he saw it, a tall building "presents to the distant observer, its skyline and broad masses of light and shade; as he approaches nearer, the large details are made out, and add to the interest of the design. The details are further enriched by details within details, the interest increasing as the observer advances." Applied to skyscrapers, this design theory explains a more sumptuous treatment of the more visible lower parts of the building and a corresponding simplicity in the removed sections of the upper facade.

At first view, John W. Root's philosophy of commercial design stands at odds with the attitudes of Sullivan and Jenney. Root forcefully advocated adapting designs to the commercial requirements of the client, the exigencies of modern technology, and the "deeper spirit of the age." He called for the "frankest possible acceptance of the commercial conditions underlying the office building"; he sought an architecture adjusted to the "age of steam, of electricity, of gas, of plumbing and sanitation."[89] Significantly, in most of Root's own buildings, attention to commercial conditions did not preclude artistically arranged and ornamented exteriors and lavishly appointed interiors. *Commercial* did not imply mean.

Root departed from Sullivan in condemning the "profusion of delicate ornament" in commercial architecture, which could only be lost "in the midst of hurrying, busy thousands of men." Such intricate delicate ornament, perhaps like some of Sullivan's patterns, was more appropriate "for the place and hour of contemplation and repose." The majority of Root's designs concentrated strong, massive, historically derived ornament in a manner that would capture the attention of hurrying crowds. At street level, at entrances, at corners and at roofs, ornament helped give the overall mass of the building, rather than individual parts, a distinctive definition; it gave the building, in Root's words, "simplicity, stability, breadth, dignity."[90]

Root's concentrated fields of ornament, his monumental entrances, and his embellished ground-story walls and lobbies clearly grappled with the skyscraper's urban street setting, a significant gesture. Photographic views presented by many critics and historians often distort this context by focusing greatest attention on the upper facades of buildings, the parts farthest removed from sidewalk encounter. Like other architects, Root designed his buildings to stand on busy downtown streets. Contemporary journalist Julian Ralph appreciated the difficulty of experiencing Chicago's skyscrapers, for in its rushing, bustling street life, all three hundred acres of downtown could be compared to the New York Stock Exchange. On streets where it seemed that "men would run over horses if the drivers were not careful," Ralph recommended hiring a cab and lying down, with face pointed upwards, to look at the great business buildings.[91] Responsiveness to the street context, to the massive presence on the skyline, and to human perception helped determine the concentrated ornamentation of entrances and lobbies as well as the overall aesthetic development of the skyscraper.

In building after building, then, Root joined other Chicago architects who opted for cultural association over structural expression. These architects showed an enduring fondness for ornamentation and a general reluctance to give architectural expression to the steel-cage structure.[92] Architects who drew on traditional forms did not foster an arid design formalism. With historicist designs, they sought to suggest continuity with past cultures that served the late nineteenth century as a familiar repository of civilized art and beauty. With patterns of foliated ornament, they sought associations of nature, a source of sensibility, refinement, and beauty counterpoised to commerce. Design for a hurrying culture was less attuned to the more abstract, subtle aesthetics of bare, structurally expressive forms, which were so often dismissed as "hitching post" or dry goods box architecture.[93]

Ornament linked skyscrapers not only to the past but to other contemporary and related structures. The tall office building did not redefine downtown alone nor develop its elaborate system of human and mechanical services in isolation. Department stores, hotels, and theaters incorporated, and in some cases established, the forms of service found in skyscrapers. Luxury hotels built in the 1820s and 1830s led the way in introducing Americans to new mechanical systems of lighting, heating, and plumbing. In their emphasis on rising standards of domestic convenience, comfort, and human services, hotels set a pattern that skyscrapers followed.[94] Hotels, theaters, and various cultural institutions in the second half of the nineteenth century shared with new office buildings sumptuous and rather elegantly appointed interiors (figs. 115, 116). Moreover, downtown department stores shared many elements with city skyscrapers. Women's parlors, libraries, and lunch clubs introduced into some office buildings in the 1890s followed department store precedents.[95] Like skyscrapers, too, department stores finally employed embellished entrances and interiors that gestured to taste and culture (fig. 117).

Even with their similarities, important distinctions existed between these related buildings and the skyscraper. The skyscraper functioned above all as a place of work. Most of the time, one entered to earn a living rather than spend leisure time or money. For members of the middle class, stores, theaters, and hotels were the site of relatively fleeting visits; office buildings became a mainstay in the daily lives of large and growing numbers of city residents.[96] Skyscrapers, then, belong within a larger ordering of the nineteenth-century city. From the middle-class view, by the close of the century Chicago comprised a variety of linked and gendered realms of life, including: first, residential enclaves, reserved for sensibility and culture and increasingly identified with consumption; second, department stores, public initiators of private consumption that facilitated respectable female presence in downtown; third, downtown skyscrapers, workplaces for white-collars and professionals, arenas for sales, advertising, and marketing; fourth, factories, warehouses, lumberyards and other sites of productive labor—a blue-collar world located along the riverbanks and rail lines and, increasingly, on city outskirts, and surrounded by working-class residences. In this context, skyscrapers were a particular piece of the city; they mediated between production and consumption, work and leisure, commerce and culture.

Yet a fair number of late nineteenth-century Americans played with the notion that a skyscraper might stand for the whole of their metropolis. Contemporaries spoke of tall office buildings as "cities unto themselves" or "a city under one roof."[97] Observers viewed the Chamber of Commerce as "a city within itself." In part, this characterization spoke to the building's impressive size and its resulting capacity to house people and activity: "There are more people doing business inside its walls than you will find in many prosperous towns, and the amount of business transacted here daily equals that done in some of the most pretentious communities in the country. Every branch of commerce and nearly every profession is represented here."[98] Size alone clearly made the skyscraper a source of pride and wonder.

Equally significant, however, these common characterizations of the skyscraper as city expressed a desire for the whole of urban life to resemble the rationalized realm of the tall building. Chicago's "famous tall buildings" embodied the city's characteristic "roar and bustle and energy," while at the same time excluding many of the city's notorious sources of physical discomfort.[99] In 1888, one observer noted that business in a high building "is remote from the noises of the street, and business may

115. Auditorium Building, north side of Congress from Michigan to Wabash, Adler and Sullivan, architects, 1886–89 (1897 photo). Chicago Historical Society, ICHi 18768.

116. Main hotel lobby in Auditorium Building, Adler and
Sullivan, architects, 1886–89 (1890 photo). Avery Library,
Columbia University.

117. Madison Street looking east from State, at right entrance
of Carson, Pirie, Scott Department Store, Louis H. Sullivan,
architect, 1898–1904 (1907 photo). Library of Congress.

be transacted within its door without interruption from such external annoyances."[100] In 1891, *Industrial Chicago,* a chronicle of Chicago development, declared that the skyscraper's "empire is the air. Creeping heavenward, it seems to reach beyond the smoke and noise of the city and beg for a place above the clouds. Comfort, cleanliness and light are within it."[101]

Distance from the street brought social comfort as well. As one Chicagoan wrote in 1892, the city was notable not only for "daring structures . . . which court the clouds" but also for the "varied life that pulsates along the thronged arteries."[102] Profiled in Sigmund Krausz's *Street Types of Chicago,* newsboys, shoe shines, ditch diggers, street preachers, street toughs, and a host of peddlers regularly plied their trades on downtown streets. Just as these street types became subjects of scrutiny and middle-class reform campaigns, the skyscraper helped separate white-collar workers from them and from lower-class street commerce in general. Viewed from skyscraper windows, in fact, people in the street seemed "like pigmies"—comfortably small, they could be rendered exotic for Krausz's book and for office tenants.[103]

The power of the skyscraper derived in part from its supposed embodiment of the progressive promise of the commercial city. Accordingly, building owners sometimes emphasized the metaphor of skyscraper as *city.* In 1892, the Masonic Temple building, the tallest office building in the world, opened with six of its twenty-two floors devoted to stores and shops, grouped around an open atrium court. The building management considered giving these floors street names instead of numbers.[104] Internalizing street and shopping activity, investors in the building sought to provide a commercial setting superior to any in the city. Admiring this "ideal plan," the *Economist* reported that it "would avoid the inconveniences of having to pass along the crowded streets by slow and uncertain means of locomotion in heat, or wet, or cold, and would present . . . attractions which could not be reached even on the street level in the magnificence of the ensemble and the grandeur of the architectural effect."[105] Limited in scale, closely controlled in design, housing a willing, paying, and largely middle-class tenantry, the skyscraper more readily approximated the supposed advantages of the city than the city did. The skyscraper appeared to be a smoothly functioning system, a city that worked.

As such, skyscrapers embodied important ideals of the nascent American city planning movement. Before Daniel H. Burnham laid out the Columbia Exposition or undertook his various urban planning projects, he spent years involved in the design and planning of skyscrapers.[106] In their combination of art, technology, human organization, beauty, cleanliness, and public amenity, Chicago's skyscrapers both anticipated and partially fulfilled the promise of the Columbian Exposition's model urban plan.[107]

In the 1880s, George Pullman commissioned architect Solon S. Beman to design both a model company town and a Chicago skyscraper. The two projects were conceptually allied. The nine-story Pullman building contained company and rental offices, stores and, more unusually, residential apartments. Through his sleeping car designs, his model town, and his downtown office and apartment building, Pullman demonstrated an unusual fascination with American domestic and sleeping arrangements. He anticipated that his railroad car company's general office staff would wish to live in the building's model apartments. Beman designed balconies for the apartments, which distinguished residential floors from office floors on the exterior facade, symbolically counteracting the collapse between residence and work. Pullman proposed providing more luxurious, comfortable, and convenient housing to his office workers than they had enjoyed in their own city dwellings. "Practical and yet beautiful," one critic commented. In its boldness, outline, and details, "it is ancient as a castle upon the Rhine, and as modern as the refined taste of the culture of the new civilization can long for."[108] Observers noted that model planning designs had been transferred from the model company town on the prairie to another focus of late nineteenth-century elite idealism—the skyscraper.[109]

The 1871 proposal for "Sky Dwellings," published in the *American Builder and Journal of Art,* called for residences to be built on the top floors of business buildings served by steam elevators. These skyscrapers suggested escape from some of the physical and social nuisances of urban life. In the upper floors of these commercial buildings, said the magazine, "the air is pure, and all the dust and discomfort of a close proximity to the streets may be avoided. The ornamentation of these rooms could be of a kind to correspond with the wealth of the inmates, and they could be supplied with every modern convenience . . . even the accessories of a garden, conservatory, or greenhouse could be secured upon the roofs, where whole families might stroll about . . . [with] a greater degree of privacy and comfort than is usually obtainable."[110] Here was the possibility that a skyscraper might include a residential enclave, complete with intimations of nature and a restricted promenade. Skyscrapers might shelter tenants from unreformed urban environs.

Mitigating the tension between commerce and culture, skyscrapers might transcend some of the cosmopolitanism of downtown. The Pullman building was "centrally located ... and yet sufficiently retired to escape much of the noise and confusion of a downtown site." It was "particularly adapted to the higher grades of office building."[111] Burnham and Root's Woman's Temple building managed to be both a "quiet retired holy place" and a "humming hive of business."[112] As tenants and visitors sought out office building restaurants, observatories, stores and offices on light courts and upper floors, they were promised seclusion and insulation not unlike the retreat of suburban residences—quiet, clean, protected, removed.[113]

Building management carefully controlled access to the building, promised discretion in the selection of tenants, and made an effort to mix tenants in an economically beneficial manner. To those unsettled by "intrusions" upon their residential and religious precincts, promoters offered reassurance. Said one: "Particular care will be taken in renting this building, and all objectionable occupations and persons will be rigidly excluded."[114] These assurances enjoyed some basis in skyscraper economics: elevators made all floors easily accessible and diminished the social heterogeneity among tenants that existed in earlier walk-up buildings. A skyscraper might present a more stable restricted enclave than many other urban settings.

This aspect of the skyscraper sparked enthusiasm but also stirred debate, for even while they fostered images of an ideal city, skyscrapers compounded urban problems. From within, the tall building's advantages seemed undeniable; from outside, a less optimistic view arose. In the 1890s, criticism of the skyscraper coalesced in restrictive legislation. The economic depression of the 1890s stopped most skyscraper development in Chicago, but a 1892 ordinance that limited building heights to 150 feet had already inhibited it. The ordinance limited heights to 125 feet on streets less than 80 feet wide and to 100 feet on streets less than 40 feet wide.

The city council, local press, and various professional and civic associations provided the forum for lengthy debate about the height of buildings. Businessmen, sanitarians, architects, engineers, realtors, builders, physicians, art critics, and numerous urban reformers all took positions on the proper form and height of downtown

buildings.[115] The debate revealed cultural ambivalence about new building technologies and about the city's physical and social ability to accommodate the skyscraper. Critics insisted that investors, in their close attention to winning tenants and seizing profits, should not be permitted to ignore the surrounding city or the need for orderly development. A model of internal coordination, the skyscraper manifested a disturbing degree of design anarchy in relation to the city plan and streetscape.

The debate over skyscrapers revolved around a series of apparent contradictions.[116] With their metal frames, large windows, huge light courts, and careful planning, skyscrapers promised light, airy, and healthy office accommodations. Critics attacked them for casting the city street into shadow, making the air stagnant and thus threatening public health. Furthermore, the burning of 1 million tons of soft coal annually for office building heating, elevator, and lighting plants produced large clouds of dark soot over downtown, belying the promise of light and clean air made in skyscraper rental brochures. Skyscrapers promised an efficient concentration of business; critics attacked them for congesting streets and making downtown a less efficient place for business. Skyscrapers incorporated a number of innovations in fireproofing; critics attacked them because their height and elevator shafts made traditional fire-fighting methods ineffective and increased the hazard of fires spreading to other parts of the city. Skyscrapers, their apologists insisted, represented a legitimate exercise of property rights; critics attacked them for inflicting damage on the property of their neighbors.

Architectural efforts to transcend the drygoods-box style of Chicago commercial building undoubtedly tempered contemporary critiques of skyscrapers. By self-consciously setting out to ennoble commerce with monumental forms, using rich materials, traditional architectural motifs, and expressions of white-collar cultivation, skyscraper architects facilitated their clients' relatively unfettered exercise of private property rights. They enjoyed substantial success at reformulating the nineteenth-century's view of commerce, casting material striving as the necessary basis of civility and creating a symbol that linked business with refinement and integrity. That success appears, perhaps, in the comfort with which some twentieth-century architects and critics contemplated buildings that allegedly expressed purely the spirit of modern commerce and industry.

Yet in the late nineteenth century, some critics continued to view skyscrapers as emblems of greed rather than of culture. No amount of tasteful embellishment could convince everyone that commercial striving yielded only benign results. Moreover, that attitudes toward skyscrapers revolved around mutually exclusive views that depended on whether one was inside the skyscraper or outside in the street highlights the limited nature of the conception of the skyscraper as an enclave. Elite landscape gardens had expanded into large public parks, but the reformed, white-collar workplaces provided by the skyscraper did not spur a broader movement for reform of the factory and other workplaces of Chicago's poorer citizens.

As it narrowed the architectural images of nineteenth-century commerce, the skyscraper did nothing to redress increasingly contentious relations between factory workers on the industrial periphery and white-collar workers and managers in the specialized and monumental center. Members of Chicago's civic and economic elite interested in the amelioration of social strife looked increasingly to public interests and public landscapes rather than to private property reforms. In the 1890s and 1900s, they turned to "City Beautiful" plans for monumental city halls, courts, municipal offices, libraries, museums, schools, and adjacent plazas. These public structures, which people theoretically shared in common, might stimulate pride in and civic identification with some broad conception of community, thus reducing the tensions of the commercial realm. This effort to establish a monumental civic landscape confronted the fact that, through their sheer size and novelty, skyscrapers had overwhelmed religious, civic, and cultural buildings once central to the city's public landscape and skyline. Whereas the city hall and county courthouse had earlier stood out in the center of downtown, now the looming Chamber of Commerce building took its place in a "massive, grand, and imposing scene."[117] Skyscrapers challenged attempts by the builders of Chicago's parks, churches, cultural and civic institutions, and even suburbs to monumentalize a life apart from Mammon. For all the power with which they suggested cultivation, skyscrapers monumentalized Mammon as never before.

Chapter 5

"Less of Pork and More of Culture"
Civic and Cultural Chicago, 1850–1905

118. Chicago City Hall-Cook County Courthouse, with 1858
third-story addition and new dome, occupying Public Square
bounded by La Salle, Randolph, Clark, and Washington, John
M. Van Osdel, architect, 1851–53 and 1858 (1860 photo).
Chicago Historical Society, ICHi 00437.

By the close of the nineteenth century, skyscraper office buildings had taken over the downtown. As critic Montgomery Schuyler put it in 1896, "It is indeed curious how the composite image of Chicago that remains in one's memory as the sum of his innumerable individual impressions is made up exclusively of the skyscrapers of the city and the dwellings of the suburbs." By contrast, there was little memorable about civic, cultural, and religious buildings in the city. Their "inferiority" in the "most characteristic dimension of altitude"—that is, their lack of height—signified that they were merely "incidental and episodical" in the larger urban context.[1] For Schuyler, Chicago's concentrated blocks of towering skyscrapers created the perception of a simple duality: downtown and suburb, *city* and *not city*. Tall office buildings, moreover, so overshadowed civic and cultural buildings as to become synonymous with the city itself. What claim could a city hall or a courthouse lay to being *civic* if it was the skyscraper that represented the city?

How to create an impressive civic and cultural landscape absorbed the energies of many city builders, architects, and critics of the late nineteenth century. In the context of downtown Chicago, this project took the form of exploring the relation between site, style, and the claims of the "public." This chapter explores such efforts by tracing the development of civic and cultural buildings from town settlement to the 1909 plan of Chicago. That plan, devised by architects Edward H. Bennett and Daniel H. Burnham, was an ambitious and sweeping effort to reassert civic preeminence. In many regards, the 1909 plan represented the grandest expression of the City Beautiful movement in America. As in other cities, in Chicago that movement promised a substantial urban transformation. City Beautiful proponents envisioned the harmonious development of the civic landscape. They offered proposals for grouping and uniting public buildings with one another and with the surrounding landscape. Generally conceived in classical Beaux Arts style, these public buildings were to provide the focal point for stately plazas and systems of embellished boulevards, radial avenues, and waterside promenades. Plans included settings for prominent public statues, fountains, and memorials.

Historians as well as contemporary observers have acknowledged that the spectacle of Chicago's World's Columbian Exposition, held in 1893, provided an important aesthetic model for many City Beautiful plans. The exposition's Court of Honor, an assemblage of ornate buildings clustered around an open court, inspired City Beautiful interest in carefully juxtaposed groupings of classically inspired buildings (see plate 6). Daniel Burnham and his advisers themselves declared that "the origin of the plan of Chicago can be traced directly to the World's Columbian Exposition. The World's Fair of 1893 was the beginning, in our day and in this country of the orderly arrangement of extensive public grounds and buildings."[2] It was a narrowly formalistic account of City Beautiful origins, one that, not incidentally, credited the creative genius of Burnham and the other architects of the fair with initiating the movement.

Burnham's creation chronicle suggested that the fair itself converted American philistines to a vision of civic art and beauty. As historians have noted, however, such an interpretation obscured the extent to which the exposition and the City Beautiful both participated in a common aesthetic and cultural movement—a movement rooted less in ideal models than in the complex patterns of late nineteenth-century urbanism. Challenging the primacy of the exposition as a source of the City Beautiful movement, these scholars have traced important contributions to City Beautiful ideals made by American park planning and by nineteenth-century campaigns for municipal art, civic improvement, and outdoor memorials.[3]

Beyond these influences, Burnham and Bennett's 1909 Chicago plan drew upon and embodied historic tensions pervading Chicago's elite nineteenth-century urban culture: the struggle to reconcile private commercial pursuits with public civic commitments. Chicagoans' discussions of civic and cultural buildings created before the plan can therefore illuminate the elements that lay behind it. Most important, the growth of skyscrapers crucially affected the ways that Chicagoans' thought about their civic landscape. City beautiful proponents aspired to reinstate a civic dominance eclipsed, in their view, by the city's commercial development. It was no accident, in other words, that the Columbian Exposition and the 1909 plan of the city, with its striking embodiment of City Beautiful aspirations, both took place in the city of skyscrapers. The tall office building challenged city builders to imagine new ways of representing civic commitments and civic authority. In this sense, the skyscraper served to focus many Chicagoans' commitment to civic monumentality.

This chapter, then, looks first at the efforts of Chicago city builders to carve out a civic realm from the city's founding through the early 1880s. It looks second at the debates and ideals aroused by four buildings of the end of the century—the Chicago Public Library, the Art Institute, a proposed new city hall, and a new federal building. In these projects, Chicago city builders developed and tested their thinking about the proper site, style, and relation of civic buildings to the rest of the city. As their thinking unfolded, they explored the problems created by a downtown dominated by skyscrapers and probed the limits of the skyscrapers' capacity to bespeak public spirit. In this sense, they confronted the question of whether it was possible to constitute a public realm from disparate if refined private interests.

The 1850s City Hall and the Federal Customhouse

Late in the nineteenth century, some architects looked back with nostalgia to the time when Chicago had enjoyed a central public square. Yet initially city residents exhibited mixed commitment to enhancing or even preserving it. The original layout of the townsite was ambiguous about city founders' intentions for the public square, a fact that later became the source of court contests.[4] The oldest surviving copies of the plat show block 39, bounded by La Salle, Randolph, Clark and Washington streets, bisected by an alley and subdivided into building lots (see fig. 3). It might be, as the Cook County board maintained in the early 1850s, that the public square was merely land assigned to the city and county for sale in order to raise revenue for public buildings and improvements. Others believed that the square was to remain public in perpetuity, either as an open park or as a site for public buildings. In the 1850s, proponents of this second view held sway; the public square provided a graceful, landscaped setting for a new combined courthouse and city hall. In its early history, then, Chicago replicated the form of civic distinction found in many new and old world county seats and cities—a public building set off by ample surrounding grounds.[5]

Boosterism influenced public consideration of what form and style Chicago's 1850s courthouse would take: "All sorts of opinions were entertained" on the project, with suggestions ranging from "an elegant and substantial structure" to an inexpensive modest building, and with proposed costs ranging from "a comparatively small amount, to a *plum*."[6] Joint committees of the city and county held meetings to consider "the vexed question of erecting buildings, somewhat commensurate with the wants, and comporting with the dignity of the city and county."[7] Vaunted notions of civic dignity and expression enjoyed extended consideration.

City builders considered two basic approaches. Proponents of a more economical design recommended building the courthouse on a lot facing the public square, which in turn would be left open. Economy could be achieved by ornamenting only the street facade of the courthouse and leaving the other exterior walls unadorned. John M. Van Osdel drew up two separate plans for the courthouse—one having the building face the square and the other with the building occupying the center of the square itself.[8] City and county committees accepted the latter, more distinctive plan in 1851, ignoring pleas for economy. The desire to breathe the airs of civic patriotism and prosperity overcame arguments that favored an open public square to serve as "the lungs of the city."[9] Rather than filling a narrow lot and rising up on its lot line, like nearly every commercial structure in the city, the courthouse would occupy the center of the public square. Set back from the street, it would be admirably framed by a pleasant landscaped setting (fig. 119). The domed building did, in fact, briefly dominate downtown. The courthouse alone occupied a full block, and it contrasted sharply with the hodgepodge architectural development of other downtown blocks (fig. 120).

Much of the grandeur of the 1851 courthouse derived from its setting, which was subsequently adorned with a wrought-iron fence, landscaping, and fountains. The building itself was not undistinguished. Costing fifty-five thousand dollars, Van Osdel's two-story courthouse had a quiet, symmetrical, classical design. Grand-order Doric pilasters ringed the facade and supported a pedimented roof. Two main entries were reached through central Roman arches that rose through both stories of the building. The courthouse roof was surmounted by a domed tower that incorporated abundant classical detail—most notably in the free-standing Corinthian columns that surrounded the dome. The stone courthouse, occupying the public square in the midst of Chicago's fledgling cityscape, represented for the city of thirty thousand an impressive gesture toward civic grandeur.

Booming population growth quickly placed new demands on government service. In Chicago, public buildings built to serve the city for a generation or more often proved inadequate after only a decade. Citizens and public officials soon called for more space and more monumental buildings that would correspond to the rising size and economic status of the city. In 1858, the courthouse roof was raised and a third story was added (fig. 118). This change preserved the landscape setting, but later additions proved less tidy.

In 1868, the Cook County board complained that the county clerk's office was a "huddled and disgraceful vestibule," the "Court of Probate is a wanderer" with "no abiding place. The Court rooms of our Chief Judiciary would disgrace the poorest county in the State. . . . The County Treasurer has not accommodations enough for a

119. Chicago City Hall-Cook County Courthouse, occupying Public Square bounded by La Salle, Randolph, Clark, and Washington, John M. Van Osdel, architect, 1851–53 (1855 photo). Chicago Historical Society, ICHi 00430.

120. Looking north along Clark Street past Public Square from Washington Street, ca. 1867. This photograph illustrates the contrasting forms of downtown blocks between the Public Square on the left and the commercial line of buildings on the right. Chicago Historical Society, ICHi 22337.

town collector."[10] The board proposed to purchase the city's half of the building and the public square. With the city relocated to another site, county offices would take over the entire structure. Chicago officials objected to abandoning the prominent and convenient public square site; however, they too needed additional room. Mayor J. B. Rice proposed a larger courthouse and city hall for the public square, a "fine marble building . . . a grand pile of which the city and county would be proud."[11]

As in earlier public building projects, a diversity of opinion prevailed as to architectural design. The *Tribune* reported that since "nearly every member" of the city and county building committees entertained "different plans for the enlargement," it was decided to advertise for building plans.[12] Architect J. G. Gindele submitted a plan for a completely new "costly and magnificent building, resembling somewhat the National Capitol," surmounted by a lofty dome. Other architects approached the problem more modestly.[13] The firm of Loring and Jenney won the commission with a plan that provided for the addition of two wings, each 100 feet by 80 feet, built in a style and material conforming to the original building.[14] As the courthouse expanded toward its lot lines and obliterated parts of the public square, the civic landscape lost something of its former distinction from the surrounding commercial development (fig. 121).

Many residents grew concerned about the building-over of the courthouse's parklike setting. Owners of abutting property found the plans "violative of the[ir] private and vested rights," since they had improved their land with an expectation of the continued existence of the landscaped public square. Now building additions would engross the park area and threaten the value of surrounding lots and buildings.[15]

In 1868, the *American Builder and Journal of Art* condemned the eclipse of the public square on aesthetic grounds. The journal could not raise much enthusiasm for the current building: "In design, the Court House offends less than most structures in the same conventional style; the material is favorable to massiveness of effect, albeit too sombre." The building's landscaped setting, however, was an attractive civic element that formed a city landmark. "So far as the present main building is concerned, it will lose whatever distinction it now enjoys, by the addition of its overgrown members. . . . The days of 'the square' are ended." *American Builder* anticipated future attempts to compensate for the loss of the public square: "Another decade will, beyond all doubt, justify the wisdom of those who favored the erection upon a new site of an edifice more adequate to the

metropolitan character and growing wants of the city, and of a material and style more attractive."[16]

The federal government also undertook a major building project in antebellum Chicago. In 1854, Congress authorized the construction of a joint customhouse, post office, and federal court building in the city. In site selection and building design, practical considerations greatly outweighed ambitions for civic expression. Secretary of the Treasury James Guthrie wrote to his Chicago agent that the building lot, preferably located on a corner, should be no less than 100 feet by 100 feet; since the Treasury imagined a 85-foot-by-60-foot structure, this would have provided a modest setback from the property line. Guthrie sought to judge the various lots offered for sale in terms of their relation to the "business parts of the city, the direction in which the city will probably increase most, whether the proposed lot is elevated or low & whether its surface is level . . . [and whether] either piling or blasting will be required to prepare the foundation for the building." He was also interested in the location of possible sites in relation to the districts occupied principally for business and those used for private residences.[17] The government viewed this question of location rather flexibly, stating that the convenience of the public and the interests of the government "jointly demand that the Post Offices, as well as the U. States Courts, should be located *near to* though not necessarily *in,* the midst of the business portion of the cities."[18] There ensued a tug of war between federal interests, concerned with economy, and Chicago interests, concerned with the possibilities for profit and civic grandeur.

The government received offers of eight sites for the customhouse. The most prominent and expensive of these sites were those located on lots facing the public square. Alexander White offered the government a lot 133 feet by 180 feet on the south side of Washington between La Salle and Clark streets for eighty thousand dollars. Opposite the courthouse, the site stood between the First Baptist and First Presbyterian churches. White said of the site, "Its eligible position for the purposes designed cannot be denied by any one at all conversant with the business part and growing importance of our city as being not only one of the best, but *the best* location in the city for *either* or *all* the purposes it is designed for. . . . If a public expression were had on the subject I hazard nothing in saying that two thirds of the entire people of our city would be in favor of the sale for that purpose."[19] Two other sites facing the public square were also offered. Their availability opened the possibility that the federal government might contribute to the

121. East and West wings of Chicago City Hall-Cook County
Courthouse, occupying Public Square bounded by La Salle,
Randolph, Clark, and Washington, Loring and Jenney, archi-
tects of wings, 1868 (1868–71 photo). Chicago Historical
Society, ICHi 22314.

creation of a central religious and civic precinct focused on the square. Such a precinct might have slowed the flight of religious institutions from Washington Street that started in the mid-1850s.

If that idea appealed to some Chicagoans, final decisions about the customhouse were made by officials in Washington, D.C. Settling for economy over display, the government purchased a less prominent site for the customhouse at the northwest corner of Dearborn and Monroe streets, two blocks south and a block east of the courthouse on the edge of a largely residential area (see fig. 43). The lot was 140 feet by 120 feet and cost the government only twenty-six thousand dollars, by far the least costly of the eight proposed sites. People interested in alternative sites prompted the city council to demand reconsideration. According to a citizens' petition, it was a "very inconvenient and unsuitable" site that would "greatly incommode a very large portion of persons having business to transact" with the government. People who supported the government's site selection demurred, insisting that the location was in the line of business district expansion and that "even at the present time the distance from the more commercial portions of the city is too trifling and inconsiderable to render it objectionable on that account." In case that argument fell flat, the petitioners offered a second, rather incompatible rationale: the "comparative *quiet*" of the site would actually benefit federal courtroom proceedings. Uncertain whether to cast the new customhouse as primarily a commercial or civic institution, Chicagoans were nonetheless clear that the building would substantially enhance the value of adjacent real estate. To confirm federal officials in their decision, landowners donated a fifteen-foot-wide strip of land bordering the selected site, land that would permit the widening of Dearborn Street without infringing upon the site itself.[20] Street widening would overcome one of the main objections to the building site—that it lacked a dignified street setting.[21]

Yet before the Chicago customhouse was even constructed, Congress legislated two major additions that pushed the building toward and then beyond the original lot line. In 1856, after reviewing the statistics of Chicago public business, the Treasury Department urged that Congress increase the size of the Chicago building from 85 feet by 60 feet to 120 feet by 60 feet. A year later, lobbied by Sen. Stephen Douglas and the Illinois delegation, Congress ignored the Treasury Department's opinion that the Chicago customhouse as previously expanded was of sufficient size and appropriated an additional two hundred thousand dollars for a customhouse

160 feet by 78 feet. This final appropriation forced the government to purchase more land on Dearborn Street, making the customhouse lot 192 feet by 135 feet.[22]

Efforts at economy evident in the selection of the Chicago customhouse site also characterized its design. The building lacked the exterior cupolas, columns, or pilasters that embellished the courthouse and other contemporary civic structures. Working in the newly established role of supervising architect to the Treasury Department, Ammi B. Young designed the customhouse in a simplified Italianate style. The design matched the plans Young used for numerous other buildings completed during a massive federal building campaign in the 1850s. The design had a quiet symmetry, balance, and refinement of detail. In rejecting the prevailing neoclassicism that characterized contemporary civic architecture, however, the building represented a clear effort by Young and the federal government to restrain the monumental ambitions of local communities that had previously manifested themselves in massive neoclassical building projects supervised by local building commissions for the federal government.[23]

Young's design was, in fact, much more akin to contemporary Italianate commercial designs. The distinctions were subtle: the federal building's large interior rooms for the post office, the court, and the customs service as opposed to the small rooms found in office blocks gave the federal building broader bay widths and floor heights than those of most commercial buildings. Despite its ample scale and proportion, the customhouse soon blended in with surrounding office blocks, several of which adopted a similar Italianate style (fig. 122). Moreover, lacking a spacious or distinctive site meant that the customhouse could easily have its setting visually intruded upon by adjacent buildings. In 1860, one Treasury official recommended the purchase of the adjoining lot on Monroe Street to prevent it from being "filled with shanties—obscuring the windows & making a nuisance." Benjamin Lombard, the owner of this land, tried to take advantage of government concern. He offered to sell part of his property to the government, promising to use his profits to build a dignified office block set at sufficient distance to ensure the admission of afternoon light into the post office, court room, and customs offices. Lombard promised to substitute stone for brick in the side wall facing the customhouse to create a harmonious architectural backdrop for the building. The merging of civic and commercial form led one publisher to combine reviews of the federal building and its neighbors in a single account of "the new marble block of Benjamin F. Lombard, a building nearly equal in size to the Post Office building, built of the same stone and finished in the best style."[24] (See fig. 123.) The review also called attention to the Reynolds block that, although built of brick, shared the Italianate style with its neighboring building.

Lacking a site comparable to the public square and occupying an ordinary lot in a commercial block forced the government to buy Lombard's land and to join in ad hoc arrangements to ensure that the buildings surrounding the customhouse would comport with its dignity. In the case of the federal building, the concern went beyond architectural symbolism into the realm of social relations. Besides promises of a fine architectural setting, the government received from Lombard assurances that his building would provide a fine social backdrop: it would "be of a character suitable for such a locality and calculated for offices which will be seeking locations near the P. office such as Ins. Co. offices—attorneys &c."[25] Here, the government carefully negotiating its relation with commercial Chicago.

Still, the customhouse handsomely complemented Chicago's midcentury architecture. The Treasury Department representative who inspected the completed building enthusiastically reported to Ammi B. Young that "the Custom Ho. is a splendid Building. The workmanship is highly creditable to the contractor. With a few unimportant exceptions it is as good as anything we have at Washington." The only complaint related to the "dead look" of the interior walls; a problem rectified by ordering the walls painted in "a delicate rose or peach bloom tint."[26] The customhouse did prompt the 1860s widening of Dearborn Street and its extension from Monroe Street south to Jackson Street. It also contributed to the southward expansion of the business district and during the 1860s, adjacent lots were filled with office blocks.

Civic structures and commercial buildings burned together in the 1871 Chicago fire. The courthouse, with its "fireproof" wings and its valuable city and county records, was completely destroyed. The walls of the customhouse were left standing but the interior was totally gutted. In the wake of the fire, both the federal and the city and county governments constructed massive new four-story buildings that filled entire blocks of the downtown grid. Their scale was unprecedented in Chicago; their monumental stone facades stretched along the four streets that bounded the blocks where they stood. There seemed little doubt that they would ponderously overpower the disparate collection of commercial buildings that were constructed on adjacent blocks. The architects who conceived these buildings in 1872 and 1873 could

not have anticipated that by the time they were completed in the 1880s numerous commercial buildings would seriously challenge their dominance of the cityscape. No downtown commercial builder assembled a site that approached the size of these public building sites. But as architects experimented with the technologies that would give skyscrapers their vertical rise, they created the basis for the eclipse of civic monumentality (fig. 124).

In August 1872, the city and county agreed that for "the convenience of the public," government offices would be rebuilt on the public square, block 39, bounded by La Salle, Randolph, Clark and Washington streets. The city would control the west half, with the county on the east half. Together the two would, however, collaborate on a design to give the building "a uniform character and appearance."[27] The decisions to concentrate and expand public offices on block 39 doomed the landscaped public square. The courthouse would stand a modest twenty feet back from the lot line; this space amounted to little more than a broadening of the sidewalk—the natural landscape features disappeared entirely. Government occupied its traditional site in Chicago, but the relation of building to site had changed dramatically.

The program of the 1872–73 architectural competition for the courthouse design specified no particular architectural style for the building. The designs submitted ranged from Greek, to Roman, to Gothic, to Palladian, to Renaissance, and included a variety of eclectic ensembles.[28] Classicism did prevail among the forty-nine competition entries. The competing architects demonstrated their clear preference for classicism in civic architecture two decades before the Columbian Exposition. Architects Wheelock and Thomas designed and described their entry in a manner typical of prevailing ideals: "The style of Architecture adopted is of a pure Italian character, such as is now used in the finest of modern European public buildings, and is generally admitted to be most suitable and expressive of the purpose for which they are intended."[29] Like numerous other entries, their design incorporated Doric, Ionic, and Corinthian columns and drew on aspects of the models provided by St. Peter's, Rome, St. Paul's, London, the Invalides and the Pantheon in Paris, and the United States Capitol. These plans firmly associated Chicago's government institutions with the Western tradition of civilization rooted in ancient Greece and Rome. Building grandly would evoke permanence, success, a flourishing civilized life, a proper use for accumulated commercial wealth, and a confidence in the future.

122. Looking south along Dearborn Street toward the Federal
Customhouse, northwest corner of Dearborn and Monroe,
Ammi B. Young, architect, 1855–60 (photo ca. 1867). The
four-story brick Italianate structure in the foreground adjacent
to the customhouse is the Reynolds Block, architect unknown.
Chicago Historical Society, ICHi 20729.

123. Federal Customhouse, northwest corner of Dearborn and
Monroe, Ammi B. Young, architect, 1855–60 (photo ca. 1866).
The Lombard Block is visible adjacent to the customhouse on
Monroe Street. Courtesy of the New-York Historical Society,
New York City.

124. South end of downtown as viewed from the Auditorium
Building tower (photo ca. 1890). To the right of center of the
federal building, the Rookery Building rises behind; the tower
of the Board of Trade is at the left; adjacent to the Board is the
lot line wall of Burnham and Root's Phenix Building, 1885–87.
Chicago Historical Society, ICHi 21795.

The full city council and county board chose J. J. Egan as the courthouse architect despite the selection, by their joint building committee, of three other architects as the winners of the competition. Egan had designed a massive four-story structure. The building shared the bold and exuberant three-dimensional character of contemporary Second Empire style. Classical motifs dominated; heavy rusticated-stone covered the basement and first floor level of the facade; Corinthian columns ringed the second and third floors and supported a heavily bracketed cornice line. The fourth floor suggested something of a mansard roof; pedimented projections at the roofline defined the entries below. Like numerous other designs proposed for the building Egan's structure incorporated an ambitious program of allegorical sculptures (fig. 125).

These virtues aside, false starts, political contention, public and private corruption, economic depression, and lengthy litigation delayed the completion of the new courthouse and city hall until 1885. The building process itself betrayed many of the vaunted ideals of the civic landscape. Constructed in the transcendent forms of classical architecture, the monumental building was to suggest civilized commitments to justice and honorable government, a dignified presence aloof from Mammon. However, boodle, corruption, and grand jury investigations overshadowed the project.[30] Reviewing the courthouse's slow progress, the *American Builder* editorialized, "Nowhere else under the sun can there be found such a ravenous and unscrupulous set of political leeches as in the city of Chicago and the opportunity to expend a few millions of dollars on a public building is one that they naturally desire to improve to the fullest extent."[31] Built to house the unified operations of local government and to symbolize a unified, prosperous community, the courthouse project ironically engendered extreme disunity. The city council struggled with the county board; Democratic and Republican "rings" battled for control; architects competed for favor (first in the public design competition and later in an alliance with different political factions); contractors sued subcontractors; and the public alternately approved and disapproved the sale of bonds for the building.[32] In 1885, Chicago had gained a "dignified" public building through an undignified process (fig. 126).

Misadventure similarly plagued construction of the new federal building. For a new Chicago customhouse and post office, Congress mandated the purchase an entire block of downtown land, either the one upon which the earlier customhouse stood or one of the twenty-four

blocks located within two blocks of the old site. The federal building would now stand alone on a full block of the Chicago grid. Owners of the block where the old customhouse stood sought a total of \$2,199,708 from the government for their lots, a price tag that aroused a "feeling of indignation and disgust" in the Chicago mayor.[33] Although the downtown area was almost completely cleared of buildings by the fire, land prices reflected prefire patterns of economic activity. Once again high land prices militated against construction of the federal building around the public square. The federal government passed by the opportunity to associate itself with the rebuilt courthouse and city hall. It moved the customhouse farther south, to the block bounded by Clark, Adams, Dearborn and Jackson streets.

Alfred B. Mullett, the Treasury Department's supervising architect, designed the new Chicago customhouse and post office. He eclectically drew upon a diverse range of European sources in designing his picturesque Second Empire monument. Pitched gables and bulbous mansarded domes topped the building's projecting bays, complementing the busy roofline of gable windows, chimneys, and turrets. The building's style and monumental scale seemed an appropriate expression of the federal government's rapidly expanding postwar powers and Chicago's rising national prominence. Poor construction, poor materials, and the apparently crushing weight of the place did not. Before the building rose much above its foundations, the walls had begun to crack. In 1875, the secretary of the treasury appointed civil engineer William Sooy Smith, architect George B. Post, and builder Orlando W. Norcross to investigate the condition of the foundation structure. The commission reported that it would be "impracticable, inadvisable, and a waste of time and money to proceed further with the construction of the building." The investigators criticized the construction and supervision of the Chicago project, expressing surprise that the "dangerous character of the foundations and the defects of the stone" were not recognized earlier. They concluded soberly, "Upon whom the grave responsibility rests for such neglect it is not the province of the commission under its instruction to inquire."[34]

Architect Gurdon P. Randall seconded the commission's conclusion, insisting that "if the work goes on it will be a mass of ruin before the building can be finished."[35] Nevertheless, a mayoral commission of local architects and engineers decided that the building could be safely completed. After a second Treasury Department investigation, Mullet's successors, W. A. Potter and

125. Chicago City Hall-Cook County Courthouse, occupying
Public Square bounded by La Salle, Randolph, Clark, and Wash-
ington, J. J. Egan, architect, 1873–85 (photo ca. 1890).
Chicago Historical Society, ICHi 19264.

126. Court interior, Chicago City Hall-Cook County Courthouse, J. J. Egan, architect, 1873–85 (photo ca. 1885–90).
Chicago Historical Society.

James B. Hill, modified the design and construction continued (fig. 127). But the contentious debates that had swirled about the city-county building now had enveloped the federal building project.

In light of these experiences, it is perhaps not surprising that some Chicagoans questioned the contemporary association of dignity and repose with heavily ornamented and expensive classical styles. Civic monumentality was so much the occasion for graft or poor workmanship that it became plausible to associate ornament with waste. At the same time, moreover, skyscraper builders were working out "businesslike" approaches to monumentality that were based on steel-frame construction. Experiments with ways to make skyscrapers signify refinement and public spirit also opened the way for a rethinking of the nature of civic monumentality. Although tall buildings posed the important problem of dwarfing civic buildings, they raised new possibilities for civic design. Some city planners looked to skyscraper technology for what it might offer to civic expression.[36]

Four projects and proposals show how the thinking of Chicagoans about the civic landscape unfolded at the end of the century. These projects served as something of a prelude to the 1909 plan of Chicago. Eventually the three completed projects, the public library, the buildings for the art museum, and the next federal building, were all viewed as possible cornerstones of a lakefront civic center and of the City Beautiful movement. These disparate projects drew upon and extended elite perceptions of the importance of distinguishing civic buildings from commercial structures.

The Chicago Public Library was one building that the 1871 fire created rather than destroyed. Believing that the fire had burned Chicago's public library, Thomas Hughes of London led a campaign in England that resulted in the donation of over seven thousand books to the city. In fact, Chicago had never supported a public library. The Hughes donation itself led to the founding of one. When it opened in 1874, the library was greeted as a welcome sign of civic beneficence and a democratic commitment to culture. Unfortunately, the strength of that commitment appeared open to question because of the peripheral quarters assigned to the library. After making an unsuccessful bid to take over and rebuild the fire-damaged customhouse structure,[37] the library occupied the third and fourth stories of a mercantile building. In the case of the library, subsequent efforts to distinguish commerce from culture had a literal dimension; they focused, in part, on removing the library collection from its commercial setting.

In 1881, Chicagoans energetically established a building fund for the public library, drawing upon private contributions. Although the project eventually failed, the effort brought to the fore many of the ideas and sentiments that underlay the late nineteenth-century civic landscape. From the outset, the library project was unmistakably the project of a particular social class. The building fund scheme originated with a Commercial Club dinner given over to the question: "Has not the time arrived when the merchants of Chicago should do something to promote the library and the art interests of Chicago?" William F. Poole, head librarian, and William M. R. French, director of the Academy of Art, both addressed the Club. Expanding his ideas a short time later, French raised the specter of Chicago's lack of "sweetness and light": "In sober truth, we are in a state of barbarism as regards these appliances of higher education." If "unconscious barbarism may be excused . . . pretentious barbarism is intolerable."[38] In a short time many of Chicago's leading businessmen were working to establish a separate library building, possibly to be connected with an art museum. Prominent citizens T. B. Blackstone, R. T. Crane, Nathaniel K. Fairbank, Marshall Field, Lyman Gage, Carter H. Harrison, Edson Keith, Franklin MacVeagh, George M. Pullman, Lambert Tree, and many others joined the effort.[39]

Civic leaders pursued the 1881 library building project as the most fitting way to mark the tenth anniversary of the Chicago fire. Competing ideas of mounting temporary pageants and street parades were dismissed as irrelevant and inappropriate ways of marking the anniversary.[40] As a purely postfire institution, a public library might symbolize a city not only rebuilt but refined. The library building was unquestionably to represent city history, and Mayor Harrison recommended that it include a special memorial hall, with a beautiful vaulted ceiling, and contain an archive of fire-related artifacts and benefactions as well as bound volumes listing the names of contributors to the building fund. The memorial hall would pass the lessons of benevolence "down to the latest generation." The building would also make an important addition to the "metropolitan equipment of the city," and place Chicago "in line with the great cities of the world."[41]

Several of Chicago's prominent religious leaders—George C. Lorimer, William Edward McLaren, William H. Ryder, Swing, Thomas—actively supported the fire memorial library project. Clergymen generally rejected the claims of some boosters that the building represented a good business investment, worth backing because it would attract visitors and residents to the community. Instead, religious leaders saw in the memorial the same

127. Chicago Federal Building, occupying block bounded by Clark, Adams, Dearborn, and Jackson; Alfred B. Mullett, William A. Potter, James B. Hill, architects, 1873–80 (photo ca. 1888). Even as the adjacent Ownings Building, southeast corner of Dearborn and Adams (Cobb and Frost, architects) overtopped the post office, it seemingly borrowed its quirky gable from the gables lining the roof of the federal building. Chicago Historical Society, ICHi 22307.

128. Stockyards (photo ca. 1920). Chicago Historical Society,
ICHi 20649.

glory of Chicago." The dome, the building, and the collection would link Chicago to the broad tradition of Western culture and would signify, in Storrs's view, a belief "in the culture of this city, in its great intellectual growth and development."[48]

In spite of its auspicious beginning, the memorial library project floundered. Enervating delays occurred because of the legal uncertainty surrounding the use of Dearborn Park as a building site. A specific building plan never evolved as a focus for the project. Organizational disputes over the fund raising system also slowed progress. Within about six months the movement died, and in 1886 the library was relocated to rooms in city hall.[49]

Finally, in the early 1890s, in the midst of continuing concerns over civic unity, cultural development, and landscape expression, the library board built a monumental building to house its expanding collection. In many ways the 1890s building project answered the aspirations of those who had energetically promoted a library building a decade earlier. When the board advertised for plans, it specified that the building was not to have a dome or tower, thereby eliminating one element of Storrs's 1881 vision. However, the requirement that the building incorporate a "classic order of architecture" and that it be built of granite or blue Bedford limestone without the brick and steel-cage construction of contemporary skyscrapers, suggested that the building was to be, in Storrs's words, "less of steers and less of pork, and more of culture."[50] The library board's specification that architects avoid skyscraper construction signaled the presence of an alternative to low, classical building. By the 1890s, adopting skyscraper technology in a cultural building was a provoking possibility.

The new library design, created by Boston architects Shepley, Rutan and Coolidge, followed the Beaux Arts classicism of two of the most impressive contemporary American library projects: McKim, Mead and White's Boston Public Library and John L. Smithmeyer's Library of Congress. Yet in spite of these important precedents, the classicism of the Chicago design (fig. 129) did not go unchallenged. Controversy centered upon the possible clash between monumental classicism and the values of economy, utility, and convenience. As William H. Jordy has noted, this tension and its possible resolution proved particularly important to American Beaux Arts and City Beautiful planning. Jordy writes that the "Beaux-Arts architect justified his decisions functionally, and . . . reaffirmed his obligation, as an architect, to dress-up the efficiencies of the present in the monumentality of the past."[51] In the case of the Chicago library, the library

129. Chicago Public Library, Michigan Avenue from Randolph to Washington streets, Shepley, Rutan and Coolidge, architects, 1891–97 (photo ca. 1899). The tower of the Montgomery Ward Building, northwest corner of Michigan and Madison, Richard E. Schmidt, architect, 1897–99, is visible farther south along Michigan Avenue. Chicago Historical Society, ICHI 22319.

board itself drew up a complete set of interior plans aiming at an economic allotment of space and at "obtaining a maximum of daylight, in conjunction with a maximum of floor space."[52] In spite of some criticism from architects who took exception to a lay board making building plans, the library board basically required that architects competing for the design dress up its efficient floor plans in a "classic order of architecture."[53]

The board received the most formidable criticism from its former head librarian, William F. Poole, who considered the plans "faulty from beginning to end."[54] For well over a decade Poole had been an outspoken critic of library design. In 1881, for example, he wrote, "Why library architecture should have been yoked to ecclesiastical architecture, and the two have been made to walk down the ages pari passu, is not obvious. . . . The same secular common sense and the same adaptation of means to ends which have built the modern grain elevator and reaper are needed for the reform of library construction."[55] As Poole saw it, new forms of construction presented designers of civic and cultural buildings an opportunity as great as their challenge.

For Poole, the most important method of adapting means to ends involved rejecting the prevailing library stack and book storage system. Poole recommended a compartmentalized library with separate reading rooms housing the collections on low book shelves, easily accessible to readers. He viewed this system as preferable to both the multistoried *cathedral* stack, with the tiers of shelves ringing a central reading room, and to the specialized stack, incorporated in the new Chicago design, which placed all books in an area completely separated from the reading room. When Poole left the Chicago Public Library to help organize Chicago's Newberry Library—a major philanthropically endowed public institution—he had succeeded in installing his compartmentalized stack and reading room system. When Newberry trustees considered placing the library on the old Newberry homestead, largely for sentimental reasons, or when they adopted Henry Ives Cobb's historicist building design and grandly set it on a site facing Washington Square, largely for aesthetic reasons, Poole did not object.[56] His critique of the new public library focused on internal arrangement but pointed to exterior architecture as indicative of underlying problems.[57] Poole, the rational arranger of books, proved especially unsympathetic to a supposed need for civic expression and monumentality.

Others too challenged the library board's commitment to classicism. In 1892, the *Tribune* published an editorial entitled: "Must the Library Building Have Classical Features?" The newspaper declared, "If classicism means squattiness or darkness, then the people want none of it." The *Tribune* also raised the issue that underlay the civic center ensembles of the City Beautiful. A low, classically styled building stood at some disadvantage when juxtaposed with Chicago's skyscrapers: "such a 'classic gem' would not look well in such a setting." The *Tribune* suggested outright that a building similar to Burnham and Root's "handsome" Rookery building, an early skyscraper, would better combine light, air, and height, and provide a form "not unworthy of a great city."[58]

In response to criticism, the chairman of the library board openly challenged those with the "poor taste" to attack classicism: "All great buildings conform to that style. It is the case in the Capitol at Washington, and the same feeling will be reflected in the construction of the new Art Institute."[59] In fact, the library building's classical motifs, based on Hadrian's Arch, its richly ornamented marble staircase and hall, and its Tiffany glass skylight and decorations helped establish an implied kinship with the classical roots of Western tradition (fig. 130). Critiques that presumed an opposition between classical architecture and practical utility ignored that fact that many late nineteenth-century public libraries were designed as far more than efficient warehouses for books. The willingness to build grandly, to build classically, marked Chicago's ascendancy. It signified that, in Emery A. Storrs's words, the city could approach the "eminence where it can do something that won't pay; won't pay in any pecuniary sense."[60]

When the new Chicago library opened in 1897, *Harper's Weekly* favorably reflected upon the building's contrast with Chicago's commercial architecture: "We have so long been accustomed to think of Chicago architecture as a rather crude embodiment of brute force, asserting itself by Brobdingnagian height and ponderousness, that it is a delightful relief to find that Messrs. Shepley, Rutan and Coolidge, the architects of the new Library have adopted the classic style."[61] *Harper's* thus casually dismissed the claims of skyscraper owners and architects that tall office buildings might express culture and refinement. The library board sought a building that would be "an honor to our city, ethically as well as architecturally, without profusion of meaningless ornament on the one hand, or of common place simplicity on the other, but aiming to convey, exteriorly, th[e] idea of dignity and repose."[62] Whatever else it might aspire to, a skyscraper could not easily represent repose; moreover, tall office buildings generally did actually *pay,* a virtue that, to

130. Book delivery station and rotunda, Chicago Public
Library, Shepley, Rutan and Coolidge, architects, 1891–97
(photo ca. 1897). The mosaic and glasswork in the room were
manufactured by Tiffany Studios. Library of Congress.

those seeking civic expression, seemed rather a vice.

The site selected for the new public library building also projected some of the same concern with civic expression. The building was built on Dearborn Park, on the west side of Michigan Avenue, between Randolph and Washington streets. The site thus conjured up for the library project the broadly public associations that characterized its park origin and use. Built as a freestanding structure on a block separated from adjacent commercial development by alleys and streets, the public library maintained an important degree of separation. Like the 1870s federal building and the city-county building, the public library almost completely filled its lot, leaving no room for a distinct landscape setting. But in facing directly onto Lake Michigan and the lakefront park, the library achieved great distinction from the adjacent sections of Chicago's commercial downtown.

Several months before winning the public library commission, Shepley, Rutan and Coolidge received the commission for another one of Chicago's leading civic buildings: the Art Institute of Chicago. The landscape and institutional histories of the library and the art museum enjoy many important parallels. Artistic and literary interests were conceived within the same cultural framework of uplift and civilization. Efforts at the monumental expression of institutional existence supported both projects. Both the library and the art museum rose from a close juxtaposition with mercantile interests to assume specialized forms—expressed in classical style and with distinct landscape settings. While the library occupied a site facing the lakefront, the museum relocated from a site on the west side of Michigan Avenue to a site on the lakefront itself, just east of Michigan Avenue. The 1891 museum project represented an important cornerstone for ambitious lakefront civic center plans of the 1890s and early 1900s.

Expositions gave Chicago art interests one of their earliest public forums and later their grandest architectural expression. The Inter-State Industrial Exposition opened in 1873 in a massive building on Chicago's lakefront. It served as an exhibition space for Chicago manufacturing interests. Some of the Exposition's stockholders decided to offer evidence of Chicago's postfire artistic progress and mounted a loan exhibition of painting and sculpture as a regular part of the annual industrial displays. Commerce and culture appeared united. From this foundation of temporary public art displays, art interests took more permanent form. In 1879, the Chicago Academy of Fine Arts was formed. Leasing space in a downtown commercial building, the academy housed an art school and an exhibition gallery.[63]

In 1882, the academy board changed the institution's name to the Art Institute of Chicago and purchased a lot at the southwest corner of Van Buren Street and Michigan Avenue. This lot, facing the lakefront, combined the visibility and accessibility of Chicago's downtown with the advantages of facing Lake Michigan somewhat removed from commercial associations. In discussing their site in 1888, the board affirmed, "The situation is beautiful, fronting full upon Lake Michigan. . . . Michigan Avenue will no doubt continue to be regulated as a boulevard, and property will retain its select character. Proximity to the heart of the city is an overwhelming recommendation in the eyes of the Trustees, for . . . convenience of access is essential." The trustees condemned other American arts institutions that had located in inaccessible, remote areas of various American cities.[64] The site chosen for the museum embodied an ideal that later civic center planners understood—the lakefront offered building sites that were in the Chicago downtown but not really part of it.

In 1885, Burnham and Root designed a massive four-story Romanesque-style structure for the Art Institute. A Belvedere style torso and numerous cameo-heads of famous artists and other sculpture on the exterior facade indicated the building's character. A Romanesque arch enclosed the door and provided entry to the building. Even with the ornamental elements of the facade, the overall structure, with its corner turrets and its formidable rock-faced exterior, suggested to one reviewer a treasure house, a safe deposit box. The building appeared "as a place where business bustle should be exchanged for respectful quiet," a place "jealous of unwarranted intrusion."[65] In a downtown area dominated by flat-topped brick buildings, Burnham and Root distinguished the Art Institute with a massive pitched gable and a heavy sandstone exterior—forms and materials more readily found in picturesque domestic neighborhoods. Although there were many "titanic and pretentious neighbors" such as the Pullman and Studebaker buildings, this overgrown art and culture house was reported to "fairly hold its own in striking and pleasing originality."[66] (See fig. 15.) According to the *Tribune*, "We shall be no longer restricted to the stock yards and the elevators as illustrations of our character as a community."[67]

In less than five years the Art Institute was abandoning its domestic Romanesque for the civic classicism of Shepley, Rutan and Coolidge. The new building and its new site on the lakefront, represented something of an heirloom of the Columbian Exposition. Several factors influenced the Institute's removal to a new building on the lakefront. The Burnham and Root building had quickly filled to overflowing. It accommodated the Institute's ambitious programs with increasing difficulty. In 1889, obscuring the architectural distinction it had earlier sought, the Institute expanded into the Studebaker building next door. Even more problematic was that the Studebaker overshadowed the museum; it cast the skylighted galleries into shadow and darkness.[68]

The Columbian Exposition offered the museum an important opportunity to occupy a more monumental building in a more monumental setting. As in the case of the earlier memorial library project, the plan for a new Art Institute emerged from discussions at the Commercial Club. The plan called for a new permanent building funded by both the Art Institute and the Columbian Exposition. The building would house the Exposition's world congresses and would house the Institute at the conclusion of the fair. The lakefront site, adjacent to downtown, answered the Art Institute's desire for central location, although it meant separating the congresses from the main site of the Exposition, seven miles south. In exchange for permission to occupy public land on the lakefront, the city council required the Art Institute to make Wednesday a free admission day in addition to its established free admission day on Saturday.[69]

The new Art Institute was monumental, classical, and set in a park. The Shepley, Rutan and Coolidge design was based on the Italian Renaissance style. The facade was ringed with Ionic- and Corinthian-order columns and pilasters. Constructed of Bedford limestone, the museum design had a rusticated first floor and smooth upper walls. The facade was enriched by a decorative entablature and cornice, panels carrying the names of artists, and niches for statuary. The plan called for building around two open courtyards and provided light, included room for expansion, and grandly stretched the facade to a breadth of 320 feet (fig. 131). Like the library plan, the Shepley design produced interior grandeur that focused on a monumental marble stairway, occupying a 2,500 square-foot skylighted space in the center of the building. Providing a precedent for attempts to establish architecturally embellished terminal points on the lakefront for Chicago's east-west street, the grand entrance closed off the vista along Adams Street, spreading the Institute's symmetry, repose, and grandeur back into the city. The new Art Institute quickly became a showcase for the city—a point of destination for visitors and residents out to make the grand tour of Chicago.[70]

In the 1890s, for the last time in the century, Chicagoans reconsidered their city hall and rebuilt the federal building. Architect Henry Ives Cobb was central to

both projects. Cobb was experienced at linking cultural concerns and architectural forms. His Romanesque-style Newberry Library in Chicago (1890–92) and his impressive English collegiate, Gothic style campus plan and buildings for the University of Chicago (1891–99) both affirmed Western traditions of beauty and culture in distinctive and monumental form. In 1894, when he proposed a design for a new Chicago city hall and Cook County courthouse, Chicago residents could well expect an equally distinctive and impressive design.

On this occasion, however, Cobb joined with George A. Fuller, a contractor and builder, to propose a highly unusual scheme for a civic edifice. Dramatically eschewing the lessons of classicism and civic monumentality presented by the Columbian Exposition, Fuller and Cobb unveiled a design for a fourteen-story skyscraper (fig. 132) to replace the overcrowded, crumbling, "foul," "unhealthy," classical style courthouse and city hall building: "The building, as laid out, is a simple straightforward office building. . . . It is a business proposition, pure and simple."[71] This was not merely a matter of architectural form but of economics. Cobb and Fuller suggested that the government rent half of the new building's floor space to private tenants. As court and public business grew, it would expand into the rental areas of the building. In the meantime, rental income would pay both the interest and the principal on the required $5 million, twenty-year bonds for construction. On the grounds of efficiency, economy, and convenience, the plan proved alluring to several members of the county board. The proposal focused public attention and debate on the contrast between civic and commercial form; it encapsulated the important struggle between skyscraper monumentality and civic expression in late nineteenth-century Chicago.

Sentiment clashed with utility and helped defeat the Cobb-Fuller plan. To many Chicagoans a skyscraper courthouse appeared practical but undignified.[72] After all, downtown a skyscraper might seem unremarkable. William Le Baron Jenney believed that the skyscraper courthouse would "rob the city of the privilege of securing a handsome monumental building—something superior in design and equal to any of the artistic works seen at the World's Fair."[73] The Cobb-Fuller design incorporated numerous classical details and, unlike most Chicago commercial buildings with their shared party walls, it would have grandly occupied an entire city block. Nevertheless, the skyscraper's commercial character prompted vigorous opposition to its use as a public building. Although both religious and cultural institutions, such as the First Methodist Church and the Auditorium Theatre, had derived substantial support from

131. Art Institute of Chicago, on Michigan Avenue lakefront at
Adams Street, Shepley, Rutan and Coolidge, architects, 1892–
93 (photo ca. 1895–1910). Chicago Historical Society, ICHi
19219.

132. Proposal for a skyscraper courthouse, Henry Ives Cobb,
architect, 1894. From *Chicago Tribune* (26 August 1894),
Sterling Library, Yale University.

commercial rents, opponents of the skyscraper court-house insisted that considerations of civic pride demanded that financial calculation merit only secondary consideration in a building designed to represent the entire community. In condemning the courthouse plan the *Tribune* declared that Chicago citizens did not want a courthouse building full of offices for lawyers, doctors, dentists, tailors, realtors, and milliners: "Surely the people would rather pay for a new building outright than to adopt a method of paying for it which might suit an impecunious town but not a wealthy county like Cook."[74] The *civis* that many Chicagoans wanted to represent publicly was not one preoccupied with rental income.

The Chicago Real Estate Board fought to defeat the Cobb-Fuller plan on similar grounds. It objected, first, to the county government competing with private building owners for office tenants, and second, to a mixing of commercial and civic forms. Board member E. A. Cummings concisely articulated a key concern with civic expression that had pervaded nineteenth-century discussions of Chicago's public architecture and landscape. Declaring that the Cobb-Fuller plan lacked beauty and special fitness, Cummings criticized it as "a monument to greed" that would "emphasize the idea long prevalent in other cities that Chicago worships at the Shrine of Mammon only."[75] Opponents roundly condemned Mammon's apparent primacy and usurpation of the civic forum—the place considered most appropriate for affirming culture, beauty, and grandeur.

Chicagoan Dwight Taylor Kennard put the issue succinctly: "Is it coming to this that our County Court House is to be drowned in the monstrosity of an enormous office building, something that can in no respect clothe itself with architectural expression. . . . Are we going to build our churches in factories, our art museums in railroad depots, our parks into beer gardens? Why do we build public buildings if it is not to inspire and raise to a higher plane the hearts and minds of the people to a higher regard for their country and respect for its laws?"[76] Although a skyscraper might improve the public image of Chicago businesses, it would sully the image of Chicago government. The widespread controversy provoked by the Cobb-Fuller plan led voters to reject the courthouse bond issue proposition. Chicago made do with its crumbling, classical style government building.

Rebuffed in his plans for the city hall and county courthouse, Henry Cobb turned his attention to a new Chicago federal building (1896–1905). The structural failures of the old customhouse and post office lent credence to arguments in favor of improved methods of construction. Continued settling of the building had cracked the foundation and the superstructure, broken plumbing, destroyed plaster walls and ceilings, and flooded various sections of the building.[77] By 1892, observers designated the structure a "disgraceful old rattletrap." The proponents of modern construction exploited this bleak history in arguing that the federal government should replace its prominent civic structure with a steel-frame building modeled on the Chicago skyscraper. As they held out against skyscraper construction, the government's architects appeared to some as "fossilized bureaucrats" who took extraordinary amounts of time and money to design derivative buildings that failed to meet functional requirements.[78] In contrast, Chicago architects working with steel-frame and terra-cotta had effected economy, efficiency, comfort, and convenience in their office buildings. The *Tribune,* although not recommending the skyscraper model, declared that the building "should at least be up to the average of the new buildings erected in the business center of the city" during the 1890s.[79]

Proposals abounded for a modern post office building constructed with a steel-frame structure and a terra-cotta exterior after the method known as *Chicago construction.* One architect, L. J. Barr, described his plans for maximizing both floor space for the post office and natural light for federal offices and courts located above the post office. Barr assumed that the building would soon be surrounded by skyscrapers on all four sides and that the best plan would be to give the post office a skyscraper plan and build it around a huge light court. The building would stand out to its lot line, giving the post office more than adequate floor space on the lower floors. The other federal offices and courts would be located in the upper floors around a light court measuring 181 feet wide and 256 feet long—by far the largest light court ever planned for a Chicago building. Civic expression in the building derived from another feature common in contemporary skyscraper design: "A grand entrance, fifty feet wide, in the middle of each front, will provide an opportunity for architectural effect and afford admission to the building from each of the four streets."[80]

Yet the Treasury Department's supervising architect remained unconvinced by such plans: "It would not be dignified to erect a steel-frame building. The government puts up heavy masonry structures and puts them up to stay."[81] Henry Cobb's entry was chosen as the most attractive, a design that, Cobb said, was decidedly modern yet ornamental, adopting classical motifs rather than modern ones. The building's most prominent feature, a

corona-topped dome, drew upon Richard Morris Hunt's design of the World's Fair's Administration Building. The dome atop the federal building design represented a grand gesture toward traditional canons of civic monumentality (fig. 133).[82] In this instance, Cobb found himself pitted against his former allies, who vigorously called for the government to build a modern *Chicago style* structure devoid of ornamental and civic gestures.

Congressional legislation authorizing the new federal building for Chicago specified that the building fill its entire site to the lot line (fig. 134). In maximizing the horizontal space available for postal operations, the legislation doomed even the modest setback of the 1870s federal building (fig. 135). Cobb designed a building that cleverly responded to the exigencies of the site and the law. He gave the post office a two-story base that covered the entire site. Above this he added four six-story wings that radiated from the central point of the site and extended to the lot lines with courtrooms and various federal offices (figs. 136, 137). The central rotunda and dome, the building's most important monumental elements, rose through and above the intersection of the wings (fig. 138). Thus the crucial domed section of the

133. Chicago Federal Building, occupying block bounded by Clark, Adams, Dearborn, and Jackson, Henry Ives Cobb, architect, 1896–1905 (photo ca. 1905). The skyscraper to the right is the Marquette Building, northwest corner of Dearborn and Adams, Holabird and Roche, architects, 1893–94. Chicago Historical Society, ICHI 18264.

134. Plan of first floor, Chicago Federal Building, Henry Ives
Cobb, architect, 1896–1905. National Archives.

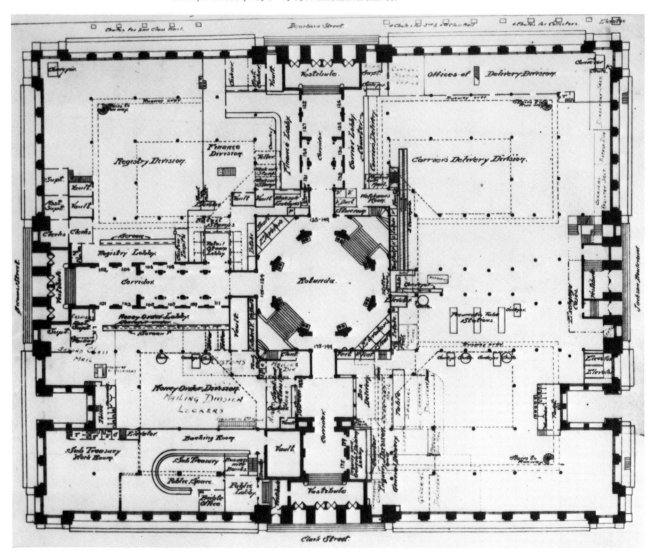

135. Plan of Federal buildings and their sites, 1860–1905, by
Areta Pawlynsky.

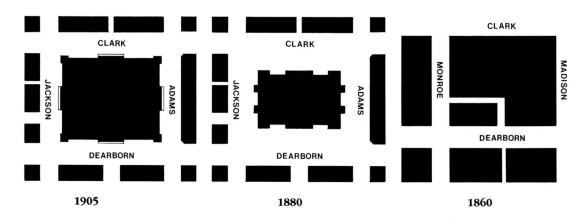

136. Plan of sixth floor, Chicago Federal Building, Henry Ives
Cobb, architect, 1896–1905. National Archives.

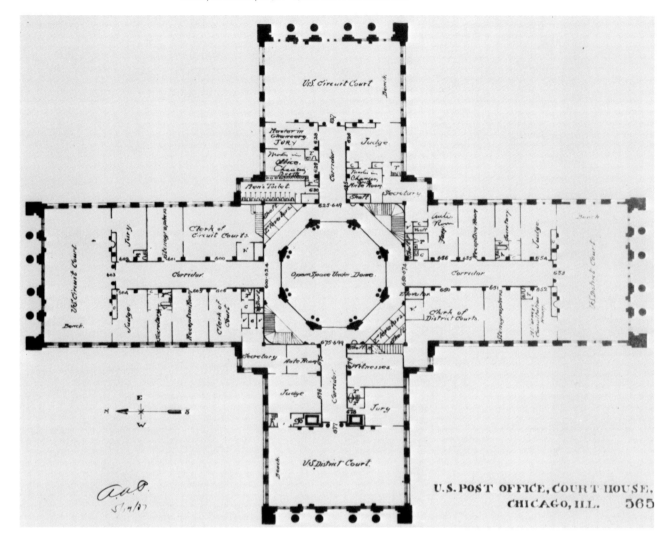

137. Courtroom, Chicago Federal Building, Henry Ives Cobb, architect, 1896–1905 (photo ca. 1905). National Archives.

138. Central rotunda and dome, Chicago Federal Building,
Henry Ives Cobb, architect, 1896–1905 (photo ca. 1905).
National Archives.

139. Looking north along Dearborn Street from Jackson, at left
is Chicago Federal Building, Henry Ives Cobb, architect, 1896–
1905 (1907 photo). Shows contrast between the massing of
commercial buildings on their lot lines and Cobb's set-back design for the federal building. At left in the foreground is the
corner of Burnham and Root's Monadnock Building, 1889–91;
at the right is Burnham and Root's Great Northern Hotel,
1890–92. Library of Congress.

building received a distinctive setback despite the building's lot-line development—a requirement that in Cobb's view worked against any "effect of foreground."[83] Although the building was only eight stories, the dome rose to 275 feet, topping such adjacent skyscrapers as the Monadnock to the south and the Marquette to the north. The cross-shaped upper portion of the building also created, in effect, exterior light courts that offered some of the same advantages of light as central skyscraper light courts. However, it did so by inverting the skyscraper light-court model and pulling the skyscraper's bustling, office-lined facade back from the street into a more removed and dignified setting. The low corners of the building stood out sharply in the midst of the dense corridors of commercial buildings along Clark, Adams, Dearborn, and Jackson streets. Here again civic form asserted the power of the setting or of the void that commercial interests, in their more intensive use of urban space, abhorred. In its context, the building demonstrated the conspicuous waste of space central to civic design (fig. 139). Full Corinthian columns supported a pedimented attic on the street fronts of the upper wings. The colossal-order pilasters on other sections of the walls suggested civic or refined purpose yet permitted more light to penetrate the interior than full columns ringing the entire building would have admitted.

Predictably, perhaps, the building came in for criticism for its "borrowed mantle of the renaissance." In 1896, Henry Lord Gay, an architect noted for his modern church interiors, objected to the traditional elements of Cobb's design, especially to the dome. He insisted that

> the first thing to be considered in a building of this character is utility. Two of the most important questions are those of light and air. . . . For a post office I do not think we should be considering domes and Michael Angelo decorations. They are well enough in their places but this building is intended for business purposes, just as the Monadnock, the Masonic Temple, or any of those splendid blocks. . . . Ask a few sensible business men if they would accept any such arrangement if they were putting their money in a building and expecting to get due return. I don't think they would.[84]

The post office stood out in the midst of Chicago's downtown; it anticipated important aspects of the City Beautiful and it opposed the alternative vision of a merged civic and commercial form. In Cobb's federal building classicism was manifest. The building's wings

and setbacks provided the dome with the hierarchically ordered setting and the distinction afforded central buildings on landscaped plazas or framed by minor buildings as featured in City Beautiful plans. All these elements seemed inappropriate to Henry Gay. Monumental civic expression was out of place in the commercial downtown: "Let the man who plans something for the Field Museum or the Lake Front go in for domes and decorations."[85] In the City Beautiful plans that culminated in the 1909, the city's designers took up this suggestion. At the turn of the century, architects of civic Chicago proposed to relocate those buildings that represented governmental and cultural authority. They were prepared to abandon the downtown to commerce.

Chapter **6**

"The Keystone of the Arch"
The Civic Lakefront
1893–1909

Jules Guerin, *Chicago: Proposed Plaza on Michigan Avenue,*
west of the Field Museum of Natural History in Grant Park,
Looking East from the Corner of Jackson Boulevard,
published as plate 128 in *Plan of Chicago,* 1909. Reproduced
from an original painting, Chicago Historical Society.

In 1896, the editors of the *Chicago Tribune* considered the architecture of Chicago's older civic buildings "radically unsuited to the business center of a great city."[1] Where skyscrapers dominated, even ornate, classic buildings, massive but low, failed to form a meaningful civic and cultural landscape. Although Chicago could boast a city hall, a federal customhouse, a public library, and an art institute, critics such as Montgomery Schuyler remained unimpressed by anything but tall office buildings. In this situation, growing numbers of architectural commentators suggested that public buildings be modeled on the commercial skyscraper and built with modern steel frames. New York architect Ernest Flagg, for example, thought it absurd to "instinctively" bypass other building types in favor of ancient models. Flagg proposed that civic buildings should "out-Herod Herod." Skyscrapers were not inherently "undignified," and Flagg predicted that public buildings would soon "be carried to such amazing heights that the tallest commercial building will be dwarfed by them."[2] For him, civic dominance needed to be literal as well as symbolic: skyscraper technology promised the way for civic structures to rise above commerce.

The City Beautiful movement in Chicago and other cities moved in a different direction; it conceded the skyline to the aggressive assertions of private and commercial interests, which competitively multiplied urban sites vertically. The power and civic presence of City Beautiful plans relied on architecturally harmonious ensembles of buildings articulated horizontally. The plans exuded a control and unity that extended from single buildings and groups of buildings outward along the axes created by new radial streets and boulevards. The City Beautiful's aesthetic unity cultivated the power of metaphor in attempting to suggest common civic purpose and shared public institutions. The aesthetics and social expressions of the plans contrasted sharply with the privatism and disunity apparent in commercial forms and activity. Commercial skyscrapers dominated the air. City Beautiful plans called for civic buildings to seize the ground.

This chapter treats a variety of civic center plans that arose in the 1890s and culminated in the 1909 plan of Chicago. These plans, which exemplified the nascent City Beautiful movement, originated among opponents of skyscraper courthouses and federal buildings and addressed the challenge to civic expression posed by tall office buildings and other disorderly aspects of the commercial cityscape. The project of conservatively restoring the distinction enjoyed by Chicago's earlier public buildings involved a continued commitment to an aesthetic of hierarchy and order. It involved, as well, clearly relating various civic buildings to one another and to spaces in-between. Most important, perhaps, these plans proposed the grouping of civic buildings in a setting that was physically and expressively detached from business pursuits and the bustle of street commerce. The 1909 plan, for example, defined a new core of the city formed around a grand boulevard that linked monumental civic and cultural centers. In this impressive reorientation of the city, the commercial district would occupy a decidedly secondary position. Much like the separate domestic realm anchored by parks, bolstered by churches, and centered in residential enclaves and eventually suburbs, the proponents of a monumental civic center conceived of a separate realm for civic and cultural life. Perhaps not surprisingly, supporters of these City Beautiful plans also envisioned them as playing a role in reshaping Chicago politics and class relations. Through spatial organization, they sought to influence consciousness and frame social interaction. In civic center plans of the turn of the century, Chicago's city building elite set out to create an arena where civic life would conform to its own idealized sense of order.

Chicagoans' various proposals for grouping public buildings to form a notable civic realm drew upon both foreign and local precedent. The European Beaux Arts tradition, Haussmann's designs for Paris, the creation of Vienna's Ringstrasse, for example, provided important models for civic monumentality. In nineteenth-century Europe, public architects contended with somewhat different problems. They worried less about competition from commercial forms than with the need to cast off the "swaddling-clothes of the Middle Ages"—removing medieval forms, such as walls around the city, that restrained development—and codifying new political and social relations.[3] Europeans sought to preserve and renew long-standing monuments by removing accretions that intruded upon the setting of what they viewed as the city's leading monuments. Elite Chicago's pretensions to European cultural forms and its aspirations to commercial empire made these models of urban redesign compelling.

Of obvious and more immediate influence on initial City Beautiful plans was the formal grandeur and classicism of the 1893 World's Columbian Exposition where a series of buildings, sharing a common cornice height and grouped around a formally designed water basin, impressed visitors as a rare example of built order and

harmony on a massive scale.[4] As a leading architectural
spectacle of late nineteenth-century America, the memory of the Columbian Exposition could be invoked to
support civic display and monumentality. Outlining his
own civic center plans, architect Normand S. Patton declared, "Imagine the Lake Front covered with a system of
buildings equal in beauty to the most attractive spot in
our memorable White City. Everybody remembers the
Court of Honor. As compared to the court of honor that
would be erected on the Lake Front the one at the
World's Fair was but a miniature."[5] The center would
make permanent the fair's transitory grandeur.

By his own account, Patton—called the "father of the
movement for the complete development of the Lake
Front," devised his civic center plans in response to
more lasting elements of the Chicago cityscape and more
endemic tensions in his culture.[6] He was galvanized by
the "disgrace" of the city's serious consideration of the
Cobb-Fuller plan for a skyscraper courthouse.[7] In 1895,
writing to Frederick Law Olmsted's firm, Patton sketched
the history of Chicago's early civic center movement:

> The whole movement was the outgrowth of a paper
> read by myself over a year ago . . . in which I protested
> the tearing down of our County Court House and
> replacing it with a building sixteen stories high . . . I
> called attention to the fact that Chicago is inferior to
> every other city in this country or in Europe in the
> beauty of its business district; that most other cities
> have one or more public squares or parks in the heart
> of the city around which are grouped the finest speci-
> mens of architecture. In Chicago we have no such
> space. . . . Therefore I called the architects to crusade
> in favor of the grouping of the public buildings in
> some place where they will face an open ground and
> can be seen to advantage. The city has such a space on
> the Lake Front and therefore our efforts have been
> directed towards utilizing a part of this Lake Front as
> the site for the future city hall, museums, and other
> buildings of a public and monumental nature.[8]

In making his argument for the lakefront, Patton looked
beyond the unseemliness of the skyscraper courthouse
to the earlier loss of civic distinction arising from the
building over of Chicago's public square (fig. 140).

In their style, conception, and underlying principles,
civic center plans responded directly to the emergence
of the city's skyscrapers. The stateliness of City Beautiful
plans depended on architectural unity, harmony, and
control. The aesthetic power of the whole center, with
its extensive plazas and patterns of hierarchy between

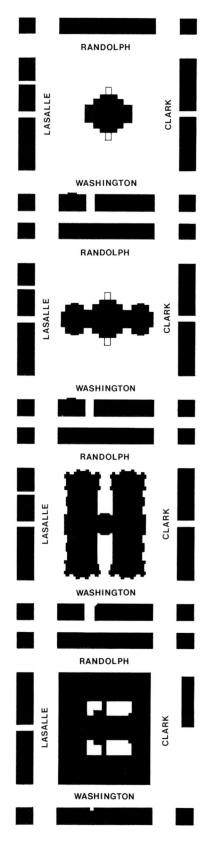

140. Public Square and the successive plans of Chicago City
Halls and Cook County Courthouses, 1853–1911. Plan by
Areta Pawlynsky.

buildings, was intended to appear greater than the sum of its individual buildings and spaces. Patton argued that Chicago's public buildings would be the "glory or disgrace of our city, according as they are erected wisely on a well-considered plan, or left to haphazard. It is not enough that such building should be beautiful in itself," he insisted; "we must have our finest buildings massed in some one place and arranged according to a preconceived scheme."[9] It was precisely this ideal of coordinated design that Chicago's commercial landscape lacked. Individual skyscrapers and commercial buildings shared common party walls with adjacent buildings, but architects made little attempt to define a shared aesthetic between the competing buildings of commercial downtown. In the 1890s, many of Chicago's downtown blocks were occupied by a diverse mixture of 1870s office blocks and towering skyscrapers that had been raised over the following two decades. The commercial downtown presented sharp contrasts of building scale, color, and materials; the civic downtown would be different.

Patton quickly moved to broaden and institutionalize efforts to distinguish and beautify Chicago's civic landscape. He established the Public Buildings and Grounds Committee within the Illinois chapter of the American Institute of Architects as a forum for objections to the skyscraper courthouse plan. In October 1894, this committee gathered a coalition of municipal reform and building interests and founded the Chicago Municipal Improvement League "to secure for our city such an arrangement, design, and adornment of our public buildings and grounds, streets, boulevards, and other public works as shall most contribute to the convenience and enjoyment of the public; [and to] stimulate an appreciation of art and give to the city a fit expression of its greatness."[10] Patton's ideas were broadly embraced, as other people also contributed important plans for the lakefront area. Alderman Martin B. Madden, the head of the City Council Finance Committee, promoted legislation to ensure that the lakefront would be devoted to public rather than commercial ends. He objected to reducing the lakefront to the "ugliness, the bustle, the turmoil and the uncleanliness" of a commercial shipping harbor, and he supported instead a civic center with "galleries, institutes, libraries, museums, and exhibition buildings . . . grouped and arrayed together more conveniently and more impressively than similar structures can be in other places in the world."[11] (See fig. 141.) Washington Porter, a wealthy Chicago fruit merchant and real estate investor, who had been a leading promoter of the World's Fair, also presented an ambitious lakefront park and civic center plan.[12] In December

1894, the Commercial Club devoted a dinner meeting to the question: "What Shall Be Done with the Lake-Front?"[13]

Daniel H. Burnham was conspicuous by his absence from the diverse group of early advocates and supporters of a lakefront civic center. Burnham's plans for the lakefront were first presented in June 1895 and October 1896 and were derivative of the Patton and Improvement League plans. Central elements of Burnham's City Beautiful ideals, such as the lakefront park and boulevard system, were prompted by input from others. In 1895, for example, when Burnham presented his plan for a lakefront civic center, Washington Porter asked him if he had considered a park, boulevard, and lagoon system for the shoreline to Jackson Park (fig. 142); Burnham replied that "the proposition was feasible, but it had not been taken into consideration in connection with the Lake Front Park." Porter sketched the plan further and underscored its potential for developing into "one of the most superb drives in the world." Burnham responded that "something of the kind can eventually be put in at both ends [of the lakefront civic center]."[14]

Within a few months of the Cobb-Fuller courthouse proposal, architects, engineers, artists, realtors, lawyers, aldermen, businessmen, and reformers were actively supporting a variety of City Beautiful plans for the Chicago lakefront. Burnham's late entry to the planning of the lakefront presents something of a historiographical problem, since many scholars have credited Burnham with originating lakefront plans.[15] Burnham himself probably promoted the idea. In 1907, his partner, Edward H. Bennett, prevailed upon one writer to change a draft of an article on Chicago planning in order to assign Burnham exclusive credit for the creative design of the lakefront park, boulevard, and lagoon network.[16] Whatever the origin of this interpretation of Chicago's development plans, however, it slights the broad contributions of the Improvement League and of figures such as Patton, Porter, and Madden. Equally important, it slights the substantial roots of the movement in Chicago soil and obscures the deep-seated tensions between civic and commercial interests reflected in civic center and City Beautiful plans.

The Lakefront

The downtown lakefront emerged as the common ground among nearly all of the diverse proposals for civic centers and City Beautiful improvements in Chicago. The site had both symbolic and practical virtues to recommend it. Important civic associations had accrued

141. Alderman Campbell's modification of Alderman Madden's
plan for the Lakefront Park and civic center, including an
armory for the First Brigade Illinois National Guard and an
Exposition building. From *Chicago Tribune* (12 May 1895),
Sterling Library, Yale University.

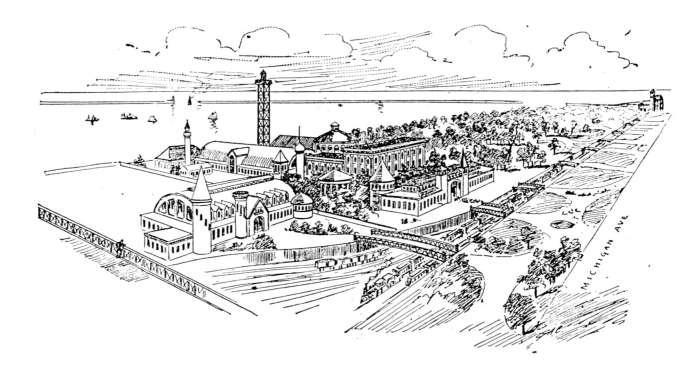

142. Chicago Architectural Club plan for South Shore Drive,
Howard Shaw, architect, Hugh M. G. Garden, delineator, 1896.
Chicago Historical Society.

to the lakefront over the years since the 1830s, when the Canal Commissioners had dedicated the area as a public ground. At midcentury, the downtown lakefront served as the central focus of public promenading and provided a cornerstone for the city's far-flung park system. The ground also served as the site of the Inter-State Industrial Exposition building. Constructed in the aftermath of the Chicago 1871 fire, that building housed annual trade shows and conventions and celebrated Chicago's post-fire reconstruction. After twenty-five years of conten-tious litigation, the public character of the lakefront was affirmed in an 1894 Supreme Court decision that set aside the Illinois Central Railroad's claim of riparian de-velopment rights. The legal victory over the "land grab" appeared as a triumph of public interest over private greed, and of civic sentiment over commercial expedi-ency.

By the 1890s, then, the area's historic development had made the lakefront seem an appropriate, highly vis-ible site upon which to affirm commitments transcend-ing commerce. The *Inland Architect* wrote, "Would it not be better to there begin an improvement for all classes of public service than to see viaducts, ware-houses, and docks shut in what in a few years will be the greatest city on the globe, if greatness be counted by heads and commercial importance? . . . How much greater will Chicago stand if her public buildings show that art has had a chance to grow and indicate a superior civilization as well as superior commercial intellect."[17]

To distinguish public buildings and space to "indicate a superior civilization" involved creating a ground apart from streets and blocks congested with commercial ac-tivity. The project required relatively clear ground. Fire had conveniently removed unwanted commercial ele-ments from the site of downtown skyscrapers; much less substantial changes would clear the lakefront. Propo-nents of lakefront plans hoped to rid the area of the Illinois Central Railroad tracks, but otherwise there were few prior improvements to contend with.[18]

Practically, as park planners had earlier noted, the lakefront possessed the most dramatic natural scenery in Chicago—welcome relief from the dreary flatness of the prairie surrounding the city. Michigan Avenue and the lakefront could provide public buildings with the conve-nience of a downtown location without the indignities of noise, dirt, and the unflattering contrasts of commercial streets and buildings. A lakefront civic center would be in downtown but not part of it. Publicly owned lakefront land had obvious advantages in a city where the cost of

assembling sprawling sites where monumental groups of public buildings could "be seen to advantage" could prove to be prohibitively expensive. Dependent on the largely horizontal expressions of classical models and City Beautiful aesthetics, early civic center plans almost naturally involved the publicly owned, open space of the lakefront, which the city planned to enlarge with landfill to a point 1,200 feet east of the existing shoreline.

The efforts of planners to turn a public ground into a City Beautiful complex shows the contrasts and con-tinuities between the park and civic center movements. Both physical and social continuities existed between the lakefront park and the lakefront civic center, be-tween the earlier ideals of the park movement and those of the City Beautiful.[19] Nevertheless, many important distinctions existed. Architecturally, the distinction involved an inversion of design approach. The formal promenades and geometric features that occupied sec-ondary positions in Chicago's earlier park plans came to dominate lakefront civic center plans. Normand Patton thought that in the civic center, "art should predominate over nature and symmetry take precedence over pictur-esqueness. . . . The buildings will not be in the park; the park will be in the inclosure of the buildings." Patton's desire to achieve grandeur dictated this approach: "The greatest architectural effects ever produced . . . have been by buildings grouped around several sides of a spacious court."[20] The formal order and classical forms of the civic center buildings would eclipse the pictur-esque lines of earlier parks. Intimations of nature would give way to symbols of civilization.

Similarly, City Beautiful planners who hoped to effect civic pageantry and order favored grand designs expand-ing the scale of the formally designed areas of promenad-ing and assembly built in earlier parks. A Municipal Improvement League civic center plan provided for pub-lic pageants and gatherings with a 10,000 seat amphi-theater, rimmed with a peristyle and topped with a triumphal arch, surrounding a field 400 feet by 1,225 feet. The plan also included an exposition building mea-suring 300 feet by 1,225 feet, an armory, a police and fire building, a city hall, and buildings for the Crear Library and Field Columbian Museum. These buildings would surround the Art Institute of Chicago. The central feature of the entire plan was an open-air music pavilion for 1,000 singers and 300 musicians, to occupy an island surrounded by large interior lagoons and Lake Michigan. A promenade, 400 feet by 250 feet, faced the pavilion and the intersections of the boulevards to the north, south, and west sides (fig. 143).[21] Burnham's lakefront plan included a similar promenade: "The Place of the

143. Municipal Improvement League Lakefront Plan, 1895. From
Chicago Tribune (10 August 1895), Sterling Library, Yale University.

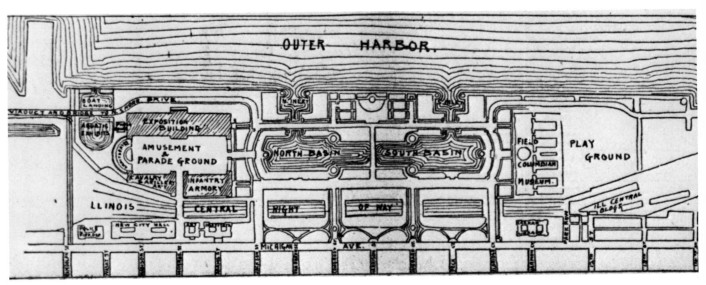

People, whereon the congestion at the center can over-flow; the place of great gathering either festive or serious in nature; the place of parades and reviews of national and civic bodies."[22] The vision of thousands of citizens gathering on public ground to share commonly held institutions guided many City Beautiful plans.

Plans did vary in their use of open space. In the Municipal Improvement League's ideal lakefront, major buildings concentrated on the north and south, and 140 acres in the center of the site remained free from buildings. Yet here the League deferred less to the supposed beauty of natural elements than to the complaints of Michigan Avenue property owners, who decried obstruction of their views of the lake. A Chicago Architectural Club plan exhibited in 1896 followed the general lines set out in the Municipal Improvement League's plan (fig. 144). Burnham made no such concession, locating the Field Museum, for which he had received the architectural commission, in the very center of his lakefront design. When questions of obstructed views arose, Burnham waved them away: "No view of a great body of water can be so beautiful as glimpses."[23] Burnham's tenacious defense of his design (fig. 145) led architect Peter B. Wight of the Municipal Improvement League to observe, "It is generally believed in this City, even among Mr. Burnham's friends—in which I think I am counted—that his personal ambition rather overweighs his artistic sense and public spirit."[24] These debates really involved matters of degree. All plans shared a basic commitment to classical buildings and formal plazas over picturesque, *natural* landscape design.

144. Chicago Architectural Club Plan for proposed Lakefront
Park and civic center, 1896. Chicago Historical Society.

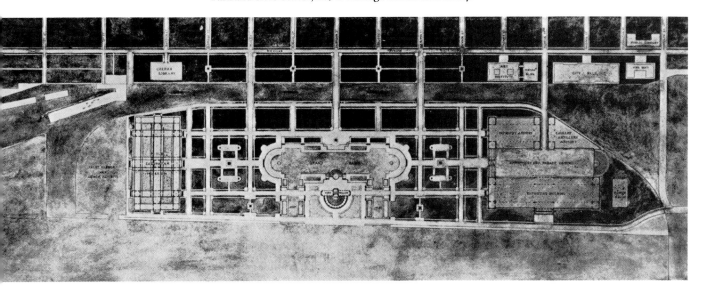

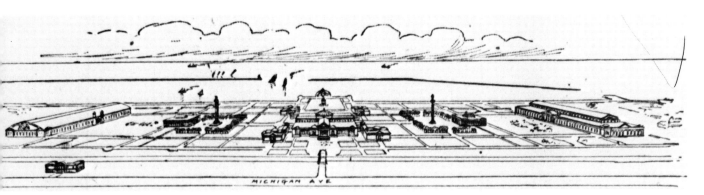

145. Burnham and Atwood plan for the lakefront, 1895. From
Chicago Tribune (4 June 1895), Sterling Library, Yale
University.

The higher degree of formality and order, the enlarging of older park promenade spaces to pervade entire plans, suggested a new order of metropolitan scale and a new order of elite social concern.[25] The park system of the late 1860s was designed for a city approaching a population of three hundred thousand people; the lakefront civic center was intended to serve a city of one and a half million. For all the emphasis on social unity in 1860s park planning rhetoric, individual consciousness and character remained a primary object of interest. Concern for the individual cultivation of gentlemen in their private gardens had expanded to take in a broader middle class at the start of the park movement. Half a century later, in City Beautiful civic center plans, the cultivation of unity among the urban masses eclipsed personal refinement as an object of focus and concern. In the civic center, an appropriate experience was not solitary reflection but keen awareness and celebration of one's membership in society.

The elite leaders of the City Beautiful emphasized the importance of a shared landscape and culture as a counter to the dangers posed by privatism and class conflict. Both obliquely and directly, they appealed to the goal of social order. Lawyer John H. Hamline, for example, suggested in 1894 that in the civic center a single image of Chicago could be presented, a cultivated image transcending the brawn and muscle of the city's commerce and industry: "I hope that in the near future there may stand out there on the Lake Front a figure of Chicago, not a nude figure indicating simply barbarity and animal force, but a figure typifying Chicago." Then pointing directly west of the lakefront and downtown, to one of the city's poorest residential sections, Hamline dramatized his call for a common landscape; here resided "a population that has in it more possibilities of damage to the future of Chicago and more possibilities of benefits to the future of Chicago than any other population in our midst. We owe it to them to make conditions so that they can look out at night and feel that they have some place where they can go and call their own."[26]

Daniel Burnham optimistically cast his civic center plan as the continuation of urban and cultural traditions rooted in Periclean Athens. The "great flowers of fine art are born on the stalk of commercial supremacy," he insisted, and his civic center would unquestionably arise from commercial profits. Yet Burnham's conservative vision did not exclude glimpses of the threat posed to Chicago by the contentiousness of the city's economic relations.

It is full time for intelligent men to turn their energies and means toward the counteraction of the growing discontent among all classes. We have wealth. Let us stop squandering it feverishly abroad. . . . Let us commence to settle down and learn to like and take care of each other, and let us make beautiful things about us, so that the happiness we seek as individuals and as a community may be helped on, so far as they affect us, by harmonious conditions. . . . The day of selfish rest either for a single man or for societies has passed away.[27]

Burnham viewed the civic center with its institutions of government, literature, art, and culture as the ideal common ground for harmonizing individual and class interests.

Alderman Martin B. Madden's plan for the lakefront openly acknowledged and then set out to reconcile the cosmopolitan, heterogeneous interests of Chicago's population. Madden sought a precise, permanent granite and marble recreation of the Columbian Exposition's Court of Honor. The entire $10 million "monument" on the downtown lakefront would be dedicated to the "rise of the Nation"; the peristyle would receive portraits and busts of distinguished businessmen, artists, and professionals. Then, in a grand catalogue of cultural diversity, the countries of the world would receive permanent exhibition space to display their "progress in civilization and their methods of religious worship." Madden proposed annual world conventions for cultural and intellectual exchange and for the "upbuilding of humanity." He hoped "to obliterate the denominational lines and stamp out bigotry and prejudice."[28] In the lakefront civic center the figure of the lone park visitor gamboling amid pastoral scenes gave way to a cosmopolitan assemblage of promenaders moving toward understanding, tolerance, and civic harmony.

Conceiving of lakefront plans in broadly social terms, people like Burnham, Hamline, Madden, and members of the Municipal Improvement League and the Commercial Club were engaged in more than abstract speculations on the Chicago polity. Rather, they addressed pressing concerns of class relations pushed to the forefront of public debate by increasingly contentious labor disputes. A bitter succession of strikes and violence rendered social relations chief among elite concerns in the twenty years after the railroad and general strikes of 1877. The Chicago police repeatedly broke up strikes by Chicago workers. Continuing labor unrest at the McCormick reaper plant culminated in the 1886 Haymarket riot, which killed seven policemen and rocked any vestige of middle-class complacency.[29] The trials, hanging, and pardons of various Haymarket figures kept labor and class issues in clear view for years to come. If, as some claimed, the Columbian Exposition engendered feelings of social harmony, such sentiments were dispelled by a bitter strike at Pullman. When President Cleveland called out federal troops to crush Pullman strikers in 1894, it became more difficult to harbor illusions of cross-class mutuality or the sufficiency of passing cultural gestures, like the fair, to knit the community together. These were only the most prominent and bloody of the labor conflicts of the era; strikes, lockouts, wage reductions, massive layoffs, and unemployment persistently marked Chicago commerce and industry.

In 1893 and 1894, just before the lakefront civic center movement, the English writer William T. Stead helped focus public discussion upon the issues of social conditions and civic unity. In mass meetings and in his book *If Christ Came to Chicago: A Plea for the Union of All Who Love in the Service of All Who Suffer,* Stead probed Chicago's darker side—the public corruption, private greed, the vice, poverty, tenement housing, and the unemployment and suffering of the working class.[30] In one mass meeting, unionist Thomas J. Morgan blasted the "crime of silence" regarding the conditions of life endured by Chicago's working class. His pointed indictment of nativists and reformers threw the meeting into a tumult. "Now the veil has been torn aside, and you members of the G.A.R., the Y.M.C.A., of your temperance societies, of your Sons of America, and Daughters of America have been able to see the skeleton in your closet." In Morgan's formulation, the lakefront figured as a locale of working-class repression rather than civic unity: "Your laboringmen may assemble peacefully on the Lake-Front begging for work, and with the strong arm of the law are driven back into their tenement houses, that the visitors who come to see the White City might not see the misery of the Garden City which built it." He bluntly conjured a picture of working-class desperation and anarchy. It was not a "fancy picture," but a day-to-day reality: "I only say it to shake you out of your false security. And, if the pleadings of Editor Stead in the name of Christ and for justice cannot shake you out, may somebody use dynamite to blow you out."[31]

As Morgan noted, reform campaigns such as temperance and Sabbath observance represented one response to the crisis, as native-born and middle-class Chicagoans sought to eradicate elements of working-class culture among the city's growing population of foreign-born immigrants.[32] In many respects, plans for the creation of a civic sphere similarly aimed at promoting conformity to so-called American values as well as enhancing civic authority. In the City Beautiful movement, social and aesthetic emphasis was placed upon order and control.[33] Civic center ensembles gained expressive power by controlling their settings; they pushed beyond their facades and building lines to spatially order the adjacent public plazas.

Equally important elements within the City Beautiful aesthetic were numerous statues and monuments conceived as focal points for civic spaces. Civic center plazas and open spaces were not designed to stand empty. Statuary and monuments, both heroic and allegorical, reinforced and restated the conservative social ideals of the City Beautiful. Freed from concern with elevator service, ventilation, book stack arrangements and the like, sculptors of public statues and monuments could annunciate most clearly the aspirations behind grander public building plans. Although they, too, were mere fragments of an idealized civic landscape, statues and monuments importantly complemented the City Beautiful's enthusiasm for civic unity, American nationalism, and the evocation of a realm of life and feeling apart from Mammon.

Like other aspects of lakefront civic center plans, monumental sculptures had been a component of civic expression in Chicago for some years. In 1887, an assembly of thousands of Chicagoans had witnessed the dedication of Augustus Saint-Gaudens's work *Lincoln, the Man* (the *Standing Lincoln*) and heard orators associate the statue with the spirit of civic education, uplift, and patriotism: "This statue will stimulate the young men of future generations to unselfish exertion in behalf of the human race, and remind all of the great struggle for National existence and personal liberty."[34] The figure of the president from Illinois, martyred on behalf of union, might command popular loyalty and evoke feelings of nationalism.

Unlike the expressive program of civic buildings, which depended strongly upon allegory, analogy, and metaphor, the expressions of heroic sculpture could settle more directly on the traits of public character it hoped to celebrate. Saint-Gaudens sculpted a strikingly natural Lincoln with few classical accessories or allegorical features. The standing figure, wearing square-toed boots and grasping the lapel of his long coat, received an extra measure of verisimilitude through the use of a life mask of Lincoln's head. Saint Gaudens's Lincoln was a recognizable and real historical figure (fig. 146).

Four years after the Lincoln statue dedication in 1891, an estimated 250,000 people gathered in another part of Lincoln Park to dedicate a statue of Ulysses S. Grant. In

146. Augustus Saint-Gaudens's "Standing Lincoln," Lincoln
Park (1888 photo). Avery Library, Columbia University.

spite of the political scandals that overshadowed Grant's presidential administration, the General's Civil War leadership made him the most popular contemporary American at the time of his death in 1885. Chicago residents, led by Gen. Joseph Stockton and Potter Palmer, were frustrated in their attempts to have the Grant tomb located in Chicago, but they carried on their commemorative designs and established an impressive Grant monument. In a public competition, five designs proposed a triumphal arch to Grant be built at the Lake Shore Drive entrance to Lincoln Park. Other designs proposed a statue of Grant standing alone, while still others called for a equestrian monument. Chicago architect Francis M. Whitehouse's plan for an equestrian statue rising from a massive granite terrace and base won the competition (see fig. 25). Sculptor Louis T. Rebisso designed a confident-looking Grant, at ease upon his horse, holding a pair of field glasses. The terrace, built overlooking the lake, served as a point of destination for promenaders and carriage riders.[35]

The multitudes attending the dedication of the Grant Monument reassured civic leaders that the commemoration of popular heroes and idealized patriotic values might help unite the community. Franklin, Grant, Lincoln, Hamilton, Logan, Washington—all memorialized in Chicago monuments—occupied places in a secular pantheon providing the basis, it was hoped, for a civic religion and coalescence. That over a hundred thousand people "from every department of life and labor" contributed to the Grant fund suggested a limited realization of this civic coalescence. As in the case of many public building dedications, the 1891 dedication ceremony for the Grant monument took the form of a civic pageant and spectacle—carefully organized and coordinated to represent itself as the assembled community. A flotilla of four hundred naval, merchant, transit, and pleasure vessels lined the waters off of Lincoln Park, all flying flags and patriotic colors. Veterans, Grand Army post members, regular army troops, representatives of civic and political organizations, and public dignitaries assembled and marched together, observed by masses of private citizens.[36]

Inspired by the Grant dedication, *Inter Ocean* wrote: "Whatever may seem to divide us when personal ambition are to be satisfied, common interests firmly reunite us, and the popular sentiment manifests itself in harmony of action when public weal or public honor bids. The people delight to glorify their heroes."[37] Here, common interests transcended the private strife that characterized commercial and industrial life. The sponsors of

monument campaigns also looked hopefully to the didactic relation between public statues and public values. The lessons of Grant's life were obvious and for many could be recalled without a statue. But a monument could "inoculate other generations with the same spirit of patriotism, valor, and devotion which gendered the events it commemorates. . . . It will ever prompt childhood to noble efforts as it discloses to their gaze the possibilities of a citizen under their Republican Institutions."[38] Public monuments reflected well upon the beneficence and high-mindedness of the present while seeking an instructive foundation for future civic life.

Between them, Lincoln and Grant symbolized the defense of a united society and a central government in the face of the claims of local and partial allegiance. These figures might have reminded the public of the spiritual and military forces that bound Americans together. At the same time, public monument campaigns included the energetic efforts of diverse ethnic communities to establish monuments to their national heroes. Immigrant groups proudly promoted monuments to such figures as Hans Christian Anderson, Harel Havlicek, Thaddeus Kosiusko, Giuseppe Garibaldi, Copernicus, Goethe, and Columbus. These monuments emphasized the contributions made to America by various foreign cultures, knitting immigrant Americans into a cosmopolitan nation rather than celebrating a continuing diversity or separateness of immigrant life.[39] Subordinated to a larger, *common* historical narrative, the traditions and values that immigrant Chicagoans commemorated might enrich American life rather than pose an alternative to it. Ethnic statues and memorials too could serve as props for civic unity.

The campaigns for public monuments prefigured the ideals pursued by civic center proponents. During the 1890s, the Municipal Improvement League proposed relieving public authorities of the need to make politically sensitive judgments concerning donations of public monuments to the city. The popular success of the sculptural program of the 1893 Columbian Exposition, undertaken by such artists as Karl Bitter, Daniel Chester French, Edward Kemeys, Frederick William MacMonnies, Edward C. Potter, Bela Lyon Pratt, and Lorado Taft, sparked further interest in civic monuments.[40] Several of the 1890s civic center plans proposed the recreation of Columbian Exposition sculptures. Burnham and Atwood's 1895 plan for the lakefront, for example, included the reconstruction of MacMonnies's *Columbian Fountain* as well as the addition of new monuments to Columbus, Grant, and Lincoln.[41] Thus early, isolated monuments and the temporary glories of the White City both anticipated statuary's integral position in turn-of-the-century plans for the civic landscape.

In 1905, Benjamin F. Ferguson, a Chicago lumber merchant, helped to secure a central role for civic sculpture with a bequest of $1 million for the construction of monuments and statues. Ferguson had resolved to establish the monument fund during his travels in European cities, where he was struck by both the impressiveness of Europe's civic adornment and the abysmal destitution of American cities in this regard. The monumental impulse that flowed from the Ferguson bequest complemented that of the City Beautiful and aimed to foster a sense of history and civic cohesion. The first Ferguson project was Lorado Taft's "Fountain of the Great Lakes," which simultaneously evoked the majesty of the region's natural scenery and the extensiveness of its commercial enterprise. Dedicating the monument in 1913, Taft expressed the sense of cultural shallowness that monumental statuary was intended to redress. "What Chicago lacks, what all our new American cities so deplorably lack is a background. Our homes, our lives are casual. We need something to give us a greater solidarity—to put a soul into our community—to make us love this place above all others. . . . Such is the value of monuments. . . . It bears its message through the ages, reaching a hand in either direction, binding together as it were the generations of men."[42] With statuary, American city builders could locate their culture as rooted in history. Controlling the past, Chicago's elite might belay the uncertainties of present and future.

Chicago artists, architects, and civic leaders presented numerous grand schemes for the application of the Ferguson fund: triumphal arches would mark the connection of the north and south park systems on the lakefront; an arch would mark the lake entrance to Lincoln Park; a replica of the 120-foot- high Greek Apollo statue, Colossus of Rhodes, would rise at the mouth of the Chicago River; arches and statues would line the boulevard system; heroic decorations and fountains would mark street intersections; a statue of music conductor Theodore Thomas would stand in Grant Park opposite Orchestra Hall; colonnades and statuary would line the lakefront; a plaza with a central column or art work as in Trafalgar Square or Place de la Concorde, would mark the boulevard link and form the nucleus for the city's most ornate buildings.[43] Although Ferguson-fund designs generally did not capture the urban grandeur of initial proposals, the seventeen major projects completed between 1913 and 1980 present some of the most prominent works of public art in Chicago.[44]

The 1909 Plan of Chicago

At the heart of the 1909 plan of Chicago, outlined in a chapter of the plan volume entitled "The Heart of Chicago," stood a dramatic effort to create a distinct civic and cultural landscape that would dominate all the other forms of the modern cityscape. Drawing upon the plans of the 1890s, Burnham and Bennett proposed a lakefront center of arts and letters with buildings for the Field Museum, an expanded Art Institute, and the Crerar Library. The center stood in imposing axial relation to a monumental civic and administrative center located at Halsted and Congress streets. Burnham and Bennett treated the center of the city as "a single composition."[45] By placing the center of arts and letters on the lakefront, by putting the civic center one and a half miles back in the city, by creating a grand landscaped boulevard connection, and by framing the ensemble's foreground with mile-long recreational piers extending into Lake Michigan, the plan would shift the visual balance of central Chicago from private to public, from the business domain to the civic realm. To divert public attention from commercial skyscrapers at the center of the city, the plan envisioned a level cornice line for business buildings and attempted to direct all eyes along the new radial roadways, converging on the soaring domes and Beaux Arts architecture of the civic and cultural centers and their sweeping Congress Street axis. The plan represented "the very embodiment of civic life."[46] (See plates 5 and 9).

In fact, there were important commercial elements to this "embodiment of civic life"; just how were these plans negotiated between commerce on the one hand, and culture and civility on the other? Chicago's Commercial Club advisers and the principal architects frequently reiterated the 1909 plan's practical, commercial basis. Radial and axial street proposals, standard features of monumental City Beautiful ensembles, received part of their justification from their supposed ability to expedite traffic and relieve congestion. Although largely derivative of earlier studies, the plan's proposals for a reconfiguration of rail terminals, for rapid transit development, for regional roadway and park systems, and its concerns with traffic and congestion, all gave the work a certain utilitarian cast.[47] In 1908, presenting a preliminary review of the plan, Charles D. Norton, the plan chairman, insisted that the study was primarily concerned with physical and commercial development. Norton noted, toward the end of his talk, that he had not used the word "beautiful" during his entire presentation.[48] Hardheaded business interests could applaud the plan for its manifest business sense.

For all that, Burnham himself frequently insisted that the crux of the reconfiguration he sought was the civic center itself. Never one to understate his case, Burnham claimed that the overall civic center would "typify the permanence of the city . . . record its history, and express its aspirations." It would be similar to the Athenian Acropolis, the Roman Forum, and Venice's St. Mark's Square.[49] Moreover, "important as is the civic center considered by itself," wrote Burnham and Bennett, "when taken in connection with this plan of Chicago it becomes the keystone of the arch. . . . The opening of Congress Street as the great central axis of the city will at once create coherence in the city plan."[50]

The civic center served not only as the visual and practical keystone of the plan; it served as the keystone to the plan's sentimental and conceptual framework. It drew on and revivified didactic and emotional traditions of the civic landscape; it reclaimed civic dominance. Set apart from the high-rise commercial buildings of downtown, surrounded by a unified set of lower classical buildings and terminating the vista of major radial streets, Burnham's civic center's main administrative building was "surmounted by a dome of impressive height to be seen and felt by the people, to whom it should stand as the symbol of civic order and unity. Rising from the plain upon which Chicago rests, its effect may be compared to that of the dome of St. Peter's at Rome."[51] The primacy of monumental and beautiful elements grounded in extracommercial considerations could hardly be masked (plate 10).

That one element should unquestionably dominate the rest seemed crucial to city planners engaged in City Beautiful projects. Notably, they selected for preeminence the building representative of local government authority. When visitors left the reconfigured city, they would remember more than skyscrapers and suburbs; their primary recollection would be Chicago Beautiful, above all else a metropolis dedicated to civic virtues.

Throughout City Beautiful rhetoric ran a clear message of the incompatibility of commerce and culture, their implicitly necessary competition for the allegiance of the city, and their inability to share a single, appropriate landscape. Like other Chicagoans before them, City Beautiful exponents disliked the indiscriminate mixing of buildings with different purposes. According to one contemporary, commercial building jeopardized the "beauty and dignity of public structures through the possibility of mingling inharmonious architecture, of making a squalid and unworthy outlook, or of destroying scale by the erection of a 'skyscraper,' or any colossal building, that would dwarf the public structures."[52]

Sharing this view, Burnham and Bennett proved especially interested in European Beaux Arts design improvements that had cleared away inharmonious urban encroachments on the sites of civic and religious monuments. Reviewing the history of city planning, the two Chicago designers lauded Haussmann's effort in Paris to open up "great spaces in order to disengage monuments of beauty and historic interest." They approved German strides in "creating about a monumental structure free room for the beholder to see the essential parts of the building"; and they quoted with admiration Napoleon Bonaparte's fantasy of urban improvement projects for London that would remove from the areas around public buildings "the mean old structures which disfigure the fine monuments."[53] The Chicago plan also pointed to several American efforts, which drew in part upon European precedent, to revive or impart civic monumentality to American cities. On many of these American undertakings, such as a 1901 Senate commission to restore the civic dignity and spatial grandeur of Pierre L'Enfant's 1791 plan for Washington, D.C., Burnham had been involved as a consultant.[54]

In most American cities, public buildings and monuments lacked the age and historic associations of their older European counterparts; they commanded considerably less reverence. Outside a few centers such as Washington, the challenge was not so much to *disengage* specific public buildings and monuments from overbuilt surroundings as to do what the 1909 plan of Chicago proposed—to create entirely new buildings and settings and, in their wake, new civic images and ideals. Grouped public buildings not only created harmonious settings, they preempted land from intrusive commercial encroachments that would later require a European style disengagement.

Commitment to the principles of the City Beautiful might involve practical opposition to commerce and impose real limits to the possibilities for its development in areas of the city set aside as a civic realm. Charles Mulford Robinson, one of the most influential City Beautiful proponents, insisted that the civic expression of public buildings would be compromised if they "were scattered about the town and lost in a wilderness of commercial structures."[55] Since public buildings "officially stand for the town," they were entitled to the best sites regardless of the passing claims of private commercial interests.[56] That government authority might occasionally intrude further on private commercial interests was implicit in some City Beautiful plans. The uniform cornice line shared among commercial buildings lining the boulevard in Burnham's Chicago plan was premised, after all, on the enactment of height limitations on commercial structures. No commercial building illustrated in Burnham's plan stood more than fourteen stories high.

Burnham, however, more than downplayed his plan's implied limitation on future commercial building. He simply omitted the many existing Chicago skyscrapers that already exceeded fourteen stories from the plan's bird's-eye views of Chicago's downtown. He wished them away. Rhetoric aside, commercial interests apparently had little to fear from the City Beautiful's crusade for the restoration of civic dominance. After all, Burnham's Commercial Club patrons were among the leaders of Chicago commerce. Burnham, as the major architect for Chicago business buildings in the late nineteenth and early twentieth centuries, had shaped the skyscraper skyline for clients in the Commercial Club and had helped plunge the civic landscape into obscurity. These builders and business leaders were not opposed to commerce. If they were displeased with the consequences of their work, they planned to undo it largely in a symbolic sense. The City Beautiful sought balance in the landscape and in the culture; its leaders wished to leave commercial interests relatively unfettered while promoting civic and cultural institutions that would suggest that commerce could support a refined and socially peaceful society.

In order to express this architecturally, the civic landscape had to expand in parallel with commercial urbanism and even regain something of its former prominence. In Edward H. Bennett's view, the civic group's regularity of lines, unity of composition, and strength of design, would permit it to vie for distinction with buildings that dominated urban skylines.[57] Burnham and Bennett based their 1904–05 plan for San Francisco on this vision. Here they proposed a new civic center ensemble: "The Architecture of the Civic Center must be vigorous in order to hold its own and dominate the exaggerated skyline of its surroundings."[58] Establishing a dominant monumentality for civic institutions did not, of course, challenge the organization and expansiveness of commercial pursuits.

Thus, the civic realm imagined by these plans did not confront Mammon so much as it claimed a separate and more impressive presence. Frequent appeals to commercial self-interest in the promotion of the City Beautiful ideal reveal the extent to which the movement, despite its architectural ideals, was not fundamentally anticommercial. Burnham was quick to aver that "beauty pays."[59] Insisting that his plan was not for the "pleasure of the rich alone," Burnham wrote that "it is to everybody's direct, living, and constant interest that the town be made attractive to a high degree in order that people may spend their money here, that business may grow and thrive among us so that every one of us shall be employed and well paid for his work."[60] Burnham's work in city planning made him particularly sensitive to public relations. He and his advisers carefully weighed the means of gaining support for plans by involving and consulting various parties, by presenting the plan itself in a form designed to sway public sentiments, and by appealing to general feelings of civic patriotism and urban boosterism.[61] A more precise analysis of benefits did not progress beyond the notion that all ships rise on an incoming tide.

Similarly, City Beautiful advocates failed to probe deeply into the actual relation between their city government and the urban masses. When they spoke of expressing or typifying *the city,* they constructed a unitary subject, a fictive Chicago that would speak with a single voice. Significantly, City Beautiful aesthetics seemed to Burnham to be fully appropriate to a seat of colonial empire, Manila. In his 1905 plan for that city, Burnham clearly annunciated the conservative social vision that informed his civic design. He took particular pains to elevate the major court buildings. He felt that the Hall of Justice represented "sentimentally and practically the highest function of civilized society." "Upon the authority of law depend the lives and property of all citizens; and the buildings which constitute the visible expression of law, its symbol of dignity and power, should be given the utmost beauty. . . . A Hall of Justice should be treated as a thing apart, a thing majestic, venerable, and sacred. It should be above all free from the clatter of commerce . . . compelling an attitude of respect if not inspiring a feeling of awe."[62] At home in Chicago, a similarly dignified or majestic civic center might overawe immigrant and native-born workers as well as impress commercial interests.

Yet to claim the respect of Chicago's immigrants and workers, city planners devised buildings to represent a government disassociated from business. In an age when growing numbers of radicalized workers saw governments to be actively in league with management, and when labor and other interests threatened to use electoral muscle to advance their own interests, elite Chicagoans understandably found it prudent to portray government as a realm insulated from self-interest; to assert that government was and ought to be so insulated. Yet if it was self-serving, such concern for transcending self-interest might still flow from genuine concerns about bourgeois culture as well as working-class unrest. Privileged Chicagoans who contemplated changing the civic landscape might sincerely want government to keep pace with commerce. In City Beautiful programs, elite Chicagoans reasserted two valued ideological tenets: First, that government might rise above self-interest; second and correspondingly, that an individual might successfully fragment his consciousness and life in order to step aside from class interest to act impartially as a citizen in the civic arena—or, for that matter, as an architectural planner of the civic landscape. Once more, to compartmentalize the city, creating separate realms for family, religion, work, leisure, culture and government, sprang from an impulse deeply rooted in nineteenth-century urban culture. Much like the reconfiguration of the religious landscape nearly half a century before, the proposed restructuring of the civic landscape expressed a desire to live apart from Mammon, to screen at least some realms of experience and consciousness from commerce. That this goal contained its own contradictions does not discount its reality.

Viewing civic centers and classical monumentality (in Burnham's words) as the keystone of the 1909 Chicago plan, approaches the essence of the plan. Ironically, the harshest historical assessments of Burnham and the City Beautiful have often stayed closer to this essential core than more sympathetic commentary. Many critics reject the City Beautiful's formalism, the supposed unthinking adoption of "foreign," classical, European forms, as unrelated to the *true* human and reform needs of American cities. Admiring twentieth-century aesthetic modernism, and distrustful of monumental settings that recalled both contemporary and historical autocracies, many historians have looked with disfavor upon what they consider to be the "pompous and autocratic" designs of the City Beautiful. James Marston Fitch writes that the City Beautiful represented "an upper-class expression of civic consciousness" with "a lot of Baron Haussmann and precious little democracy in these vast geometries of befountained plazas and intersecting boulevards"; that all in all, City Beautiful designs were "actually peripheral to the central fact—the crisis of the city itself." Sigfried Giedion views the World's Columbian Exposition and the subsequent American City Beautiful movement as "a quite unnecessary national inferiority complex." Such dismissive interpretations overlook the broad cultural importance of City Beautiful expression: its allusions to ancient historical tradition in a city with a shallow history, its conservative effort to contend with economic conflict and civic unity, and its apology for commerce through recourse to civic culture.[63]

Notwithstanding modern discomfort with or misunderstanding of the aesthetics and cultural striving of the 1909 plan, the formidable nature of the plan and its publication has made it difficult to overlook. Some historians have salvaged the plan from historical obscurity by approaching it rather obliquely. They sympathetically laud the plan's "modern," twentieth-century understanding of urbanism, its "regional" scope, its network of railroads and highways, and its attention to freight and traffic movement. These elements, juxtaposed with traditional City Beautiful forms, suggested a more sophisticated, comprehensive regional view of the city and its interrelated parts. In tracing Burnham's transition from "piecemeal" to "comprehensive" planning projects, Cynthia Field, for example, notes that "regional planning was a concept introduced in its modern form by Burnham. His later plans, those for San Francisco and Chicago, extended the scope of planning from the traditional center city into the suburbs and even beyond. In those plans, the impact of the growing urban area on the entire region was first suggested."[64] Carl W. Condit further maintains that the 1909 plan of Chicago was "the first to be predicated on an understanding of the unity of the city and its metropolitan context."[65] Heralding Burnham as "modern" casts the nineteenth century as a dark ages of planning and overlooks many obvious nineteenth-century antecedents for Burnham's regionalism;[66] equally, it ignores the keystone of the plan.

Civic and business leaders understood the dynamic relation between Chicago and its region from the town's earliest settlement in the 1830s. The city grew in large part through the successful speculation on and planning of regional development. Commercial interests vigorously established regional road, canal, and railroad systems to develop the hinterland and promote Chicago's growth. Similarly, the dynamics of intraurban development— the link between transportation, residential settlement, and city and suburban patterns—provided the basis for action long before Burnham's plan. Burnham's regional view of the *interlocking* nature of transit, sewer and water lines, and commercial and residential forms shared a great deal with the insights of nineteenth-century builders and residential tract promoters. Extensive railroad and highway networks were projections of the belt-line and by-pass rail systems and park boulevards proposed for Chicago from the 1870s onward.[67] Extensive provision for drives and roads along the lakefront had been proposed in the 1850s and 1860s; they assumed the regional scope of the 1909 plan in the 1870s and 1880s.[68]

Radiating from monumental civic and cultural centers, the 1909 plan's modern road system was determined in large measure by the aesthetic requirement of axial approaches to groups of civic and cultural building. These roads reinforced attempts to shift urban imagery from commercial to cultural form. Burnham's contributions as a publicist and his desire for effecting political coalitions—not his vision as a planner—led him and his collaborators to incorporate these contemporary planning initiatives into the 1909 plan. Moreover, *regional* aspects of the plan were largely peripheral to its central monumental and aesthetic ideals.

Skyscrapers and Civic Buildings

Chicago's twentieth-century city builders largely ignored the civic ideals and forms essential to the 1909 plan. The bid to create civic and cultural buildings to dominate the "composite image of Chicago" ended in failure.[69] City Beautiful plans languished and the social and aesthetic aspirations behind them collapsed. The lakefront park, later named Grant Park, did indeed develop on a monumental scale during the 1910s and 1920s, as the shallow waters adjacent to the old public ground were filled and landscaped. It also provided a link toward the realization of the impressive vision of nineteenth-century park planners for an extensive lakefront park and parkway system. However, Grant Park developed as only a fragment of the 1890s civic center ideal and of the center of arts and letters proposed in the 1909 plan. Although improved by extensive plantings, architecturally embellished terraces, and several handsome statues and fountains, the lakefront park never accommodated the civic and cultural institutions envisioned for it. Legal challenges by property owners along Michigan Avenue prevented the construction of all buildings on the lakefront. As a result, the provision of additional park lands along the lakefront simply extended the plans for parks dating from the 1860s and 1870s. It bypassed the more substantive civic and cultural ideals of City Beautiful advocates.

The Field Columbian Museum, proposed by Burnham and others as the central building for the lakefront cultural center, took up a secondary site facing the south end of Grant Park. In the 1910s and 1920s, Soldiers Field Stadium, the John G. Shedd Aquarium, and the Max Adler Planetarium came to occupy spacious park sites adjacent to the Field Museum. Completely separated from other buildings, this cluster of institutions did create something of a cultural center. The Field Museum terminated an important vista along the newly developed Lake Shore Drive while the Aquarium ended the vista along Roosevelt Road. The grouping of these institutions provided each individual building with an added measure of distinction, in the general spirit of 1890s civic center plans; however, the rather informal relation of the buildings to each other and the peripheral, asymmetrical location of the entire group in relation to downtown made the complex fall far short of earlier City Beautiful aspirations.

The monumental civic center planned for an area one mile southwest of Chicago's business center never materialized. From the outset, critics thought its proposed location was an "inconvenience" to people who would have to do business there.[70] More damaging was that, even as various civic center plans were being promoted, the most important and appropriate buildings for such centers were being constructed elsewhere.[71] At the height of 1890s lakefront planning, the federal government resolved to build its new post office and courthouse on its old downtown site. In 1905, just as discussions that led to the 1909 plan were beginning, Chicago's city and county governments decided to build a new combined courthouse and city hall on its old block 39 site.

These decisions resulted partly from a change in official attitudes toward civic designs evidenced between the 1896 design for the federal building and the 1905 design for the city-county building. While Henry Ives Cobb's federal building showed some commitment to distinctively civic architectural elements, the city-county project moved toward a conflation of civic and commercial form. Unmoved by City Beautiful oratory, some Chicagoans came to believe that a skyscraper might appropriately house civic functions. Ironically, perhaps, the very monumentality of government prompted a realignment of perceptions. Under the impetus of progressive reform, municipal and county government had grown larger and was more closely analogized with business enterprises. Crusading for government efficiency, reformers praised business models; the acceptance of business buildings for government use reflected this new attitude.

The editorial page of the *Tribune* dramatized the transition. In the 1890s, *Tribune* editors had bluntly condemned the "undignified" Cobb-Fuller plan for a skyscraper courthouse. A decade later its position had shifted:

> A courthouse is primarily a "public utility." It is a place where business of great importance to the community is carried on—where suits are tried, taxes paid, and real estate records kept. A "monumental" building is not required for these purposes, and there should be no attempt to inflict one upon the city. . . .

There are enough monuments in Chicago, and the County commissioners should not harken to the ambitious architects whose minds run on pillars, domes, and turrets, but go on to provide a business building for a business community.[72]

Claiming to speak for the voters, the *Tribune* announced that Chicagoans wanted a modern office building, of steel-frame construction, for their new courthouse. Any interest in the restoration of the "glory that was Greece and the grandeur that was Rome" should be subordinated to the provision of abundant light and air and improved sanitary and elevator service. As for the proponents of beauty and sentiment: "Let us hoist the esthetes with their own petard. Ruskin says that beauty springs out of usefulness. Make the new county building useful and a clever architect will find some way of making it simultaneously presentable."[73]

This decline of interest in civic expression was part of an ebbing of civic idealism after 1900. In political discussions of the new century, critics increasingly judged local government by its utility rather than its sentiment. Starting in 1902, for example, Chicago residents and politicians engaged in a protracted debate over reforming the city's municipal charter. The charter from the state of Illinois had led to numerous inefficiencies brought on by overlapping administrative and fiscal authorities, a decentralized and divided provision of services, and severe restrictions on the ability to tax, issue bonds, and undertake a variety of municipal improvements. The charter movement adopted a tenet of progressive reform, calling for an efficient government to provide the services needed by the city's businesses and residents.[74] The *Tribune* strongly supported charter reform. In advocating a modern business building for the city hall and courthouse, it settled on a form that fulfilled its new vision of the largely utilitarian function of government.

A similarly noticeable shift in civic idealism occurred in some of Chicago's leading reform organizations. Chicago's elite increasingly turned the management of charitable and cultural institutions over to professionals trained in the delivery of service, as opposed to lay people more committed to an abstract notion of social and cultural uplift.[75] The change in the *objects* of the Civic Federation of Chicago charted something of this transition. Founded in the wake of William Stead's 1893 mass meetings to expose and reform social conditions, the Civic Federation undertook an ambitious program. It sought "honesty, efficiency and economy" in the local government and thus the "highest welfare of its citizens."

It set out to promote agencies that would "discover and correct abuses in municipal affairs" and to insulate local politics from state and national parties. These objects remained in force well after the turn of the century. However, the Civic Federation also hoped to "serve as a medium of acquaintance and sympathy between persons who reside in the different parts of the city, who pursue different vocations, who are by birth of different nationalities, who profess different creeds or no creed, who for any of these reasons are unknown to each other, but who nevertheless, have similar interests in the well-being of Chicago."[76]

The original statement of objects could easily have served as an introduction for many lakefront civic center plans. However, in 1903 the Civic Federation recast its stated objects: "The purposes of this Federation shall be local municipal improvement and the betterment of civic conditions, the promotion of efficiency in the public service, and the furtherance of wholesome legislation."[77] Gone were specific notions of civic coalescence achieved across class, race, and ethnic lines. The federation's goals had settled more firmly on municipal efficiency. Architectural debates concerning the form for the courthouse and city hall followed a similar course over the same period. Civic architecture was increasingly measured by a commercial standard, just as municipal functions were measured by their efficiency at furthering business interests.[78]

In April 1905, the competition announcement for the design of the new Chicago courthouse and city hall came just weeks after voters approved a $5 million bond issue for the building. Although the competition program did not specify a form or style for the building, Chicago newspapers had printed the general views of County Board President Brundage and his Courthouse Citizens Committee on the most suitable plan. Brundage favored a ten- to twelve-story building constructed up to the lot line, in the shape of a "commercial structure," which would include "enough variation to give it some distinguishing feature that shall mark it as a public building."[79] The majority of the thirteen competition entries accordingly proposed a large office building constructed around an interior light court. The distinguishing feature generally took the form of colossal-order classical columns adorning one or another part of the building. This facade motif, however, was already in place on some Chicago skyscrapers constructed by the city's major banks and corporations. The most novel design echoed Cobb's federal building and post office design of 1896. The design internalized, within a single building plan, the

City Beautiful's concern with spatial hierarchy, visual control and harmony, and orderly relations between disparate parts. It included a three-story base covering the entire site—entered through large ornamental arches. Three separate, five-story towers rose above the base, but were set well back from its edge. The design's effectiveness and monumentality derived from its control of an entire block of land combined with the coordinated design of the three office block units on a single block—a pattern unprecedented among competitive constructions of the commercial downtown. Another architect presented a distinguishing feature by calling for an elaborate sculptural and decorative program, while yet another offered a plan for a building topped with a forest of cupolas and turrets, like a Turkish mosque.[80]

The courthouse competition jury, headed by Massachusetts Institute of Technology architecture professor William R. Ware, bypassed these more unusual entries. It awarded prizes to designs that added colossal-order columns to the exterior of buildings with the structural form of contemporary commercial office buildings. The jury did not give its unmixed blessing to any one design. It awarded first prize to Barnett, Haynes and Burnett, whose plans appeared closest to the specifications of the competition. It awarded second place to the Holabird and Roche design, which presented the best interior features for what the jury considered the county's true requirements; and third place to Shepley, Rutan and Coolidge, whose design presented the "most attractive" exterior. The jury, in fact, recommended that Shepley and Holabird collaborate on a final design.[81]

In late August 1905, the county board exercised its prerogative to bypass the jury and awarded the courthouse commission to Holabird and Roche who, respectfully declining to collaborate with Shepley, preceded to implement their own design (fig. 147). They called for a three-story base supporting a middle section of six stories ringed by ninety-four-foot colossal-order Corinthian columns that carried a formal entablature, concealing a tenth floor. A central lobby, 21 feet high and 31 feet wide, ran across the entire building block and exceeded the interior monumentality of most contemporary office buildings. A skeletal steel frame supported the building, which was laid out around central light courts, as was familiar in office building design. Even as the building assumed its modern commercial form and plan, county officials sought to suggest that they had really struck a balance between modernists and monumentalists (fig. 148). The official county guide to the new building described it as "the first attempt in America to combine in

one structure the distinctive features of the public edifice and the modern office building, preserving the dignity of the former and retaining the utility of the later."[82] It was a combination to which Patton and the civic center designers of the 1890s would have pointed with civic shame rather than pride.

Applied to the site, a standard of efficiency led the county board to prefer the central location and convenience of block 39 over the lakefront area. Transit lines had grown up around the old courthouse, as had commercial office buildings that housed law firms and other businesses that used the courthouse. The interests linked to the old courthouse location could present formidable opposition had relocation been seriously considered. In 1905, reflecting on the vicissitudes of civic design and the courthouse site, the *Tribune* reported, "The present site cannot be improved upon. For that reason it will be necessary to sacrifice beauty to utility. . . . The chief public buildings of a great municipality like this should be a delight to the eye. That is impossible here. Nearly all the ground will have to be covered by a high building."[83] Some modernists viewed the courthouse as a prototype for public buildings constructed in the dense centers of major cities where high land prices and the size of government operations forced public buildings to "emulate the skyscrapers."[84]

Architects and planners proposed a number of plans to mitigate the modern vicissitudes of the courthouse's central location. Although already officially committed to "comprehensive" plans for the future of Chicago, Burnham, Bennett, and the Commercial Club advisers on the 1909 plan of Chicago proved most eager to participate in the piecemeal planning of the present. This was true in the case of the city hall and courthouse project. John G. Shedd, vice-president of Marshall Field and Company, member of the 1905 Courthouse Citizens Committee and architectural jury, and adviser on the 1909 plan, attempted to persuade Chicago officials not to link the city hall to the courthouse building. Instead, Shedd proposed that the city abandon its half of the public square and relocate to the block just west of the square. By placing the city hall on the western section of the proposed block, Chicago would gain a grand plaza, bisected by La Salle Street and flanked on either side by the harmoniously developed facades of the separate city hall and courthouse buildings. The two public buildings would stand opposite each other, controlling the space between and creating a block of order amid the discordant downtown district. A less desirable alternative plan was also outlined: half-block squares to the east of Clark

147. Chicago City Hall-Cook County Courthouse, occupying
Public Square bounded by La Salle, Randolph, Clark, and Washington, Holabird and Roche, architects, 1905–11 (1911
photo). Chicago Historical Society.

charitable transcendence of Mammon and the ideals of unity that they encouraged in their own congregational and parochial communities. Episcopal Bishop McLaren declared that the memorial project "seems to say: there is something higher in life than acquisition of money. . . . Notwithstanding the individualizing tendencies of our time, we are all in some sense members one of another."[42] Reverend Thomas echoed McLaren's sentiments but, as was his practice, cast them in a more secular light: "We want something that will give us the pride of a city, and the unity of citizenship in the city. . . . Nothing would do this so well as something that would be above sectarianism, and above party, and above nationality; something that would be cosmopolitan, taking us all in."[43]

Boosters of Chicago's civic and cultural life thus incorporated ideals of unity, benevolence, and transcendence of self-interest that many wanted the religious landscape to embody. Religious critics wondered if the city's material progress was matched by a "parallel moral power."[44] Similarly, in both public and private meetings, speakers and planners focused on the contrast between Chicago's material prosperity and its *cultural* poverty. The point was valid in its own terms: Chicago did lack many of the cultural institutions that distinguished cities in the eastern United States and in Europe. At the same time, it seems likely that this complaint addressed the often unacknowledged but dramatic contrast between the material prosperity of their own class and the material poverty of others. Indeed, elite proponents of the building fund spoke movingly of the importance of unifying the community, and that phrase resounded with several meanings. The claim that material success had been achieved but found wanting could be raised, after all, only with particular groups of city residents, members of the successful classes. Equally, the search for a unifying culture would take place for the most part around elite values, literature, and history. The library would provide evidence of past charity and strength, celebrate present philanthropic responsibility, and educate future generations as to the importance of these civic virtues. At the same time, many no doubt sought a unity beyond the felt limitations of contracted religious and domestic spheres. Beneficence and philanthropy were virtues intended to embrace a broader public.

Emery A. Storrs, a Chicago lawyer noted for his eloquent oratory and his humor, helped crystallize support for the memorial library project at a mass public meeting in March 1881. Speaking to a crowd that overflowed the Central Music Hall, Storrs outlined what he saw as the leading justification for commercial pursuit and its relation to local cultural enterprise. Simply stated, he felt that commerce could provide the foundation for a more civilized community life. Earlier, Storrs had linked Chicago cultural ascendancy with commercial prosperity: "The time is not far distant when [Chicago's] steers and hogs, its corn and wheat and lumber shall return to it in rather finer and more aesthetic forms; and from the stockyards and pork-packing establishments there are sure to come some day great libraries and splendid galleries, and music and the drama will find these exceedingly practical and unromantic sources their best temples furnished."[45] (See fig. 128.) In the library meeting Storrs criticized inaction and urged the audience onward:

> The time has passed when the City of Chicago can plead infancy, business pursuits, or press of other business engagements as a defense for the total neglect of anything that looks in the direction of intellectual culture. I am tired of the uniformity of its brag. . . . I am tired of being continually reminded of the vastness of the stockyards, of the extent of the grain trade, of the magnitude of our lumber interests, and of the enormous development of the pork trade in this great commercial metropolis. I want less of steers and less of pork and more of culture. . . . I want Chicago to rise to that eminence where it can do something that won't pay; won't pay in any pecuniary sense, but will pay in the larger, broader, and grander, and better sense.[46]

Storrs defended wealth even while requiring it to justify itself with culture. Appealing to Chicagoans outside a small circle of business leaders, Storrs hoped to stir civic pride and forge unity around a common commitment to the cultivated ideal. "We have not been making cultured men and women here," he said, "but we have been preparing. . . . Books will do this. Art will do this, great libraries will do this."[47]

When Storrs envisioned his cultural aspirations architecturally, he thought of Renaissance design. He saw a freestanding, domed building housing the high culture of the city—an image that reappeared in later City Beautiful plans and in the 1909 plan of Chicago. Storrs envisioned a "great building that shall face the sun on the shores of this inland sea, the shining dome of which shall greet it morning after morning, and shall salute the setting sun good night for all ages to come, and in which shall be stored the best works, the best thoughts and the best pictures of the world. This will commemorate the

148. Construction of the Chicago City Hall-Cook County
Courthouse, occupying Public Square bounded by La Salle,
Randolph, Clark, and Washington, Holabird and Roche,
architects, 1905–11 (1907 photo). In this construction
photograph, it is possible to see the grid of the structural
steel framework and the start of the colossal-order columns,
which contributed a note of civic distinction to the building.
Chicago Historical Society, ICHi DN-4898.

Street and to the west of La Salle Street would be developed as public plazas facing two sides of a combined city hall and courthouse building.

Edward J. Brundage, former president of the Cook County board and sponsor of the courthouse plan, wrote in his capacity as the Chicago corporation counsel that Shedd's plan for a plaza separating the city hall and the courthouse was "thoroughly impractical." It would force the county to spend an additional $1½ million to adorn the west facade of the courthouse, the facade left exposed by the removal of the city hall. It would also require that the city make "lavish expenditure" in condemning land that it could not really afford. Brundage concluded that future funds for beautifying the city might be devoted to creating squares to the east and west of the city and county building; "this would give the necessary distance to render a good view of the building, and I think would be far more practical than the plan now proposed."[85] Such pragmatic considerations and such financial constraints did not bode well for the fulfillment of the much more grandiose ideas for civic and cultural centers included in the 1909 plan.

Facing burgeoning calls for modernism in civic architecture and the government's decisions to scatter Chicago's public buildings around downtown, Burnham and his collaborators on the 1909 plan looked hopefully to the future. They argued that the civic and cultural centers would require a generation to build and might be realized if they were properly planned for. However, by the time government officials undertook new building plans, in the 1950s and 1960s, the sanguine assumptions of the 1909 plan for downtown expansion had lapsed. Later planners conceived of the new building projects, in part, as a way of shoring up the future of the existing downtown's office and retail structures. A civic center that would break the established pattern of connections between private and public office space and existing transit lines proved politically and economically untenable.

The Beaux Arts classicism of the early twentieth century continued to collapse the distinction between civic and commercial form. Classical style skyscrapers of the 1910s and 1920s, including such notable buildings as the Continental and Commercial and the People's Gas buildings in Chicago, designed by Burnham's firm, continued a commercial appropriation of the refined images of civic architecture. This architectural borrowing also extended beyond the skyscraper. In 1896, a two-story Burnham-designed building was presented in terms familiar from the descriptions of many civic projects:

Chicago, having given the skyscraper to the world, with its manifold conveniences and economy of operation, now presents an architectural monument worthy of any age in which art was honored. . . . After passing rows of modern office buildings where vast fronts of plate glass intrude upon the sidewalk, imagine the effect of a building of solid granite, massive and grand in its classic simplicity, a building whose mass and dignity will cause the surrounding "skyscrapers" to look thin and weak. . . . The front will have a magnificent colonnade of eight monolithic Corinthian columns, thirty-six feet in height, surmounted by a noble cornice and attic.[86]

This contrast of the modern skyscraper and classically styled lowrise had formed the paradigm of distinction between civic and commercial form in 1890s' debates over the skyscraper courthouse. But this Burnham design was not a public building; it was the new home for the Illinois Trust and Saving Bank, built on a corner that seemed to demand not a two-story building but another skyscraper (see fig. 89). As public builders struggled along with ambitious but generally unrealized plans for civic and cultural centers, private builders aggressively adopted elements of these plans and incorporated them into commercial projects.

By the mid-twentieth century, architects gave government buildings strikingly modern forms that extended even further the conflation of civic and commercial architecture evident in earlier Beaux Arts business buildings. Mies van der Rohe's 1960s design for Chicago's federal center appropriated for the government forms that had been widely used in modern skyscraper design of the 1950s. The center grouped a thirty-story courthouse, a forty-five story federal office building, and a one-story post office around an open plaza. Ironically, the federal center came to its plan of placing a building on an open plaza—the historic form of the civic landscape—not from civic precedent but rather from the leading examples of modern commercial building. The federal center and the thirty-one story Cook County civic center both brought new meaning to the traditional form of monumental plazas. In the context of 1960s Chicago, open plazas in the center of a densely built downtown drew upon the powerful associations of the federal government's urban renewal programs and its central role as the shaper of cities. Nevertheless, these civic architectural forms were nearly indistinguishable from leading examples of contemporary commercial development.

Without the monumental center of arts and letters, axially connected along Congress Street to the civic center, with its administrative building's dome dominating the skyline, the essence of the 1909 plan was gone. What Montgomery Schuyler noted of Chicago in 1896 remained true; civic, cultural, and religious buildings continued to appear "incidental and episodical."[87] Great additions to Chicago parks, especially along the lakefront, and fine but generally isolated monuments of civic, cultural, and religious life did affirm substantial commitments to matters beyond Mammon; however, as mere fragments of the grand visions of the City Beautiful movement, these monuments could hardly distract from or compete with the towering skyscrapers of Chicago business or the sprawling manufacturing plants of Chicago industry. These features of the cityscape fixed and perpetuated the image of a city given over to commercial calculation.

Burnham and the Commercial Club advisers failed where they most wanted to succeed, in adding to the commercial city a refined, uplifting civic landscape that would foster social unity. American urban government's limited powers in the early twentieth century to sell bonds, condemn land, and finance grand civic improvements severely compromised City Beautiful prospects. In the twentieth century, as commercial interests introduced novel and increasingly monumental forms, the City Beautiful foundered on the shoals of high downtown land costs, relatively low building budgets, and government unwillingness to restrict private land in the interest of municipal beauty, order, and civic expression. In crystallizing cultural attitudes toward commerce and urban form, the 1909 plan of Chicago conservatively looked back to the nineteenth century rather than forward to twentieth century's modern forms and institutions. It failed to envision the socially cohesive power of modern commercialized mass culture that ultimately played a more central role in Chicago society than elite standards of civic life and form. The failure of the 1909 plan and the City Beautiful generally reflected the ambiguity of its vision, a contradictory ideal that embodied both a critique of and a rationale for commerce.

Epilogue

In its architectural and cultural development, Chicago stood out among nineteenth-century American cities, but it did not stand alone. Urban centers in many regions of the country similarly engaged in city building campaigns to establish park systems; create specialized skyscraper downtowns; relocate religious institutions into outlying residential enclaves and suburbs; and promote plans for a monumental civic landscape. Some of the professional designers who worked in Chicago—Burnham and Olmsted, for example—maintained practices that were national in scope and that shaped eastern, western and midwestern cities alike. Furthermore, Chicago's elite did not hold a monopoly on ideological preoccupations with the effects of commerce on individual character and social order.

Though it shared landscape and architectural forms and cultural presuppositions with other cities, Chicago still stood out. The combination of a tabula rasa for a townsite, extraordinary growth, and massive economic resources concentrated greater city building energy in Chicago than in nearly any other American city. The sheer intensity and novelty of city building in Chicago compelled intense interest from outsiders and intense self-consciousness from many residents. From the outset, settlers enjoyed a palpable sense of the possibilities in founding and defining a new city.

Within a few decades, to comprehend their city, Chicagoans needed persistently to multiply population numbers and other statistics that measured expansion, as well as take account of a rapidly changing cityscape. The question of what the city would become, aside from a commercial entrepôt, persisted because few people could ignore that a great city was indeed rising on a site that some city residents still recalled as an open prairie. The city's second coming in the wake of the 1871 fire reinforced the self-consciousness of residents about the built city. By contrast, even the most rapidly expanding eastern cities had eighteenth- or seventeenth-century origins. There, landscape and social relations changed dramatically but still reflected a greater inheritance and ballast from the past than in Chicago. One small comparison makes this point: while many Chicago churches abandoned the downtown area, New York City's Trinity Church remained rooted at the foot of Wall Street. Surely the history of religious worship on Trinity Church's site, marked by the graves in the churchyard cemetery, meant that a good deal more sentiment accrued to this site than to the sites of Chicago's various church buildings. In addition, New York's Trinity profited from old investments in real estate, which insulated it from the financial pressures felt by the newer Chicago congregations.

205

Like Chicago, other newly established nineteenth-century cities also lacked influential preexisting architectural and social forms; they contrasted, however, with Chicago in the scale of their development. Among the newer cities, Chicago captured people's imagination by winning decisively the competition for urban commerce and population. Simply put, the city poured more resources into its own construction than other new cities. Chicago builders raised more and bigger skyscrapers, parks, and churches and, in the process, enjoyed a greater opportunity to develop an architecture of refinement and cultivation. The resources devoted to city building were also evident in the foundation of a thriving artistic culture in the city. Impressive elements of Chicago's cityscape were matched by high cultural productions and collections of notable scope and quality. Chicago, then, helped to define the terms of cosmopolitanism emulated by other cities.[1]

At the same time, self-conscious city building in Chicago arose from more than rapid growth. Outsiders, impressed with the city's commercial possibilities, noted shortcomings in the organization of cultural institutions and found the landscape "any thing but inviting." In 1849, one prospective settler confided to his diary that "as a young man without any encumbrances I would as soon select this spot as any other to seek my fortune—but as a residence there is little congenial to my feelings."[2] Other critics broadcast their views. In 1853, English traveler Fredrika Bremer published her account of the United States, including a brief description of the city on Lake Michigan: "Chicago is one of the most miserable and ugly cities which I have yet seen in America, and is very little deserving of its name, 'Queen of the Lake;' for, sitting there on the shore of the lake in wretched dishabille, she resembles rather a huckstress than a queen. . . . And it seems as if, on all hands, people come here merely to trade, to make money, and not to live." Nonetheless, Chicago residents included "some of the most agreeable and delightful people" Bremer had met anywhere. They were "not horribly pleased with themselves and their world, and their city." They were people "who see deficiencies and can speak of them properly, and can bear to hear others speak of them also."[3] So Chicagoans, she felt, might possibly set out to improve their city and themselves.

After mid-century, changes in architecture and urban design revised travelers' and even residents' views of the city. More benign assessments characterized the travel and booster literature of the late 1800s. Baedecker's United States guide, for example, cautioned readers that "great injustice is done to Chicago by those who represent it as wholly given over to the worship of Mammon, as it compares favorably with many American cities in the efforts it has made to beautify itself by the creation of parks and boulevards and in its encouragement of education and liberal arts."[4] The accomplishments of city builders were acknowledged even in the reform literature aimed at exposing the city's evils. One such chronicle, George Wharton James's 1891 study, *Chicago's Dark Places,* opened with a figure of an urban tourist gathering a wholly salutary view of the city. The tourist rides in an elegant carriage through the city's parks, along its boulevards and past its "palatial residences." He visits the skyscrapers, hotels, and clubs and attends a performance at the Auditorium Theatre. When he leaves the city the tourist is "full of rhetorical enthusiasm over the great and glorious—the young and beautiful city he has left behind. Its homes are 'super par excellence,'—its Auditorium, unrivalled—its parks, exquisite, its lake-view, sublime—its avenues, delightful—its future, glorious. Words fail him to express the feeling of astonishment that overcame him as he saw how the Goddess of Plenty had poured forth her golden stores into the lap of this phoenix of American cities." James's tourist did not lack cosmopolitan experience or sophisticated taste, but he had seen only part of Chicago "and, as far as he had seen, his judgment would have been correct, his enthusiasm easy to understand, and his laudation to be expected."[5] For Chicagoans had created beautiful and cultivated landscapes indeed.

The problem was that the city's transformation was partial and, moreover, that tourists were not the only people capable of viewing Chicago without seeing "the darker, the sadder side." For in the city "ordinary Chicago humanity" often tended to "look on the side of the prosperity, progress, magnificence and splendor" of the city. Chicagoans did this to such an extent that James and other reformers felt compelled to "draw aside the veil" that distorted many people's perceptions, to expose the ills of the saloon, prostitution, poverty, and poor housing.[6] City builders achieved great success in defining a series of urban landscapes that permitted them to live comfortably in Chicago separated, as if by a veil, from the less congenial aspects of the city.

The skyscraper nicely encapsulates this mode of urban perception; it purified images of commerce by providing a new, refined workplace for white-collar workers in the midst of an increasingly beautiful downtown. The skyscraper's success in reorienting images of commerce turned not only on its actual design and embellishment but on its pattern of exclusion, its abstract and distant relation—its often physically unrepresented relation—with industrial production. What the skyscraper did for images of commerce and work, a series of distinct urban places did for the city as a whole. Residential enclaves and suburbs, with their allied church buildings, together with parks, civic buildings, and monuments provided refined settings for the unfolding of a comfortable cosmopolitan existence. All of these settings worked as the skyscraper worked; they excluded or disguised, in varying degrees, industrial production and immigrant and working-class people and their cultures; thus, as in synecdoche, particular parts of the city came to represent the whole.

In the course of the century, Chicago's commentators suggested various elements to represent the whole city. Chicago might be a city in a garden, a city of churches, a city of skyscrapers, or a city of civic consciousness, the lakefront serving as its front yard. City builders painstakingly devised plans for removing a railroad track from the lakefront, while immense railroad yards, lumberyards, and grain elevators stood not far distant, and while reformers could point to befouled parts of Chicago's landscape—its dirty streets, its polluted river, the wooden tenements that clustered west of the river and throughout the stockyard neighborhood to the south and that housed Irish, Germans, Slovaks, Poles, Italians, and Lithuanians in turn.[7]

Wealthy residents could travel in a circumscribed physical and social world with only fleeting contacts with such scenes. Landscaped boulevards, lakefront drives, and commuter rail lines directly linked the more refined settings of the downtown skyscrapers, department stores, and civic and cultural buildings with outlying and elite residential areas. As one guide to late-nineteenth-century Chicago reported, "respectable people" did not go to the slums, and tourists were encouraged not to linger amidst the "hastily constructed tenements," pawn shops, gin mills, brothels, and cheap cellar lodging houses.[8] Respectable Chicagoans kept their distance from these parts of the city and, reformers complained, they did not let their imaginations wander there either.

Chicago's elite often tended to rely on the abstraction of numbers and statistics in presenting the broader city to others. Figures on population growth could hardly fail to impress, and a simple recitation of population called to mind grand scale without necessarily leading observers to focus on the miles of unpaved, unsewered streets,

on the primitive conditions of housing, and on the mile upon mile of blighted landscape. Similarly, boosters presented production statistics for the Chicago industries of national importance with great flourish, like cattle slaughtering (15,755 in 1872–73; 2,469,373 in 1892–93.) The numbers often masked a bleak reality. In 1862, reporting that Chicago "never was so flourishing," William McCormick wrote to his brother Cyrus, "Packers are here from all quarters & have built fine stone & brick buildings down on S. branch. The river is positively red with blood under the Rush St. bridge & down past our factory! What pestilence may result from it I don't know."[9]

Visions of Chicago depended in part upon where one stood among the economic classes but also upon where one stood and traveled in the city itself. Only small, bounded areas amid the vast expanse of Chicago's 211 square miles provided the focus of elite city building efforts. Thus, descriptive literature on Chicago revealed alternately one of the most beautiful of American cities and one of the most ugly. This fact led James Muirhead to characterize Chicago in 1898 as a "City of Contrasts." Similarly, in 1929 when Harvey Zorbaugh titled his sociological study of Chicago *Gold Coast and Slum,* he captured and made vivid a pattern of segregation rooted in Chicago's nineteenth-century city building that was at once both cultural and architectural.[10]

This book has disputed a conventional view of nineteenth-century Chicago as a city that reflected the profit motive but little else. The specter of mindlessly narrow materialism "uneasily entered" many people's minds; the concern for promoting an urban world beyond the marketplace significantly shaped enduring elements of nineteenth-century architecture and urbanism, a public landscape that we continue to occupy familiarly today. Yet the city described in these pages was as fully a product of capital as the city described by Carl Sandburg, Lewis Mumford, and other observers. This is not to underscore the purely pecuniary benefits secured by promoting cultivated landscapes within the city—although these were substantial and city boosters knew it. It is rather to suggest that genuine aspirations to genteel existence existed in a context of social class, with all the contradictions that entails.

City builders were not insincere (in fact, they were often obsessed with their sincerity). A critique of Chicago city builders would maintain instead not so much that they were sometimes greedy, but that they were often self-absorbed and blinkered, fundamentally more concerned with their own inner states than with outer realities. What they represented to themselves as their finest qualities—their sensibility, their refinement and taste—had deeply problematic, invidious consequences for their society. They could not imagine a way of living above Mammon that did not require the buoyancy provided by wealth. They presumed a human nature in which, in Louis Sullivan's words, self-interest and self-love were unalterably "the lower and fiercer passions."[11] It followed that human beings—like the cityscape—could be cultivated or refined but not transformed. In their preoccupation with having their virtues reflected physically in objects, landscapes, and buildings surrounding them, they took possession of certain parts of the city, imprinted them with their values and tastes, and committed themselves to a circumscribed world.

The city building campaigns that embodied this strategy did not proceed consistently or evenly. Nineteenth-century developments therefore have to be viewed chronologically. Those parks and churches conceived and promoted as keystones of a specialized, residential realm were created before the emergence of a downtown filled with skyscrapers. It was after 1880 that some skyscraper builders worked on the premise that downtown could itself provide the locus of refinement. The evidence from the history of civic Chicago is similarly nuanced. The rejection of a skyscraper city hall and the unfolding of a City Beautiful program once again asserted the importance of architectural distinction and landscape specialization. These developments in the civic landscape ambiguously suggested both the success and failure of efforts to create a refined urban setting with the skyscraper. Opponents of the proposed skyscraper city hall objected that the form was inherently undignified. Yet the elaborate City Beautiful proposals for Chicago civic and cultural centers set themselves in competition not with a tawdry commercial landscape but rather with commercial forms of monumentality and beauty that had set earlier civic buildings and their setting into broad eclipse.

The simultaneous building of a high-rise city hall and courthouse and the collapse of central elements of the 1909 Chicago plan raise the question of what happened to the cultivated ideal in its relation to city building during the twentieth century. There is little question that this ideal provided much less impetus for twentieth-century planning than it had for nineteenth-century design. This has not meant, however, that the architectural and cultural forms of the nineteenth-century have lost relevance in the twentieth century. Even with changed patterns of use and popularity, nineteenth-century park

systems remain important features of the urban landscape today. They have even been supplemented by massive regional systems. Relocated churches and the specialized domestic landscape have presaged the persistent and booming suburban development in American cities from the late nineteenth century to the present. The critical reception of twentieth-century skyscrapers has continued to engage the artfulness, taste, and aesthetics of commerce. Chicago and other American municipal governments have openly encouraged the construction of skyscrapers as a way to promote visual images of economic prosperity. Many cities have accomplished this even as their industrial bases and populations have steeply declined; to some extent skyscrapers have grown even more important as visual proxies for urban commerce. The rise of private automobiles has powerfully reshaped twentieth-century cities; it has also reinforced nineteenth-century forms of the segmented landscape by providing private links between disparate sections of the city settled and developed for refined work, residence, and leisure pursuits.

The cultivated ideal has declined as a motivating force in twentieth-century city building. The decline has paralleled the decline of the "cultivated gentlemen" who presided over deliberations concerning parks, churches, skyscrapers, and civic building. Even in the wholly public plans for parks and civic buildings, in the nineteenth century the figure of the gentleman as the proponent and recipient of these improvements loomed large. One can look far and wide in the history of twentieth-century city building and fail to locate the central figure of this gentleman. He was most certainly swamped by the burgeoning professionalism and science of city building and planning. The move from City Beautiful to city practical or city scientific ideals certainly influenced Chicago planning and design.[12] Planning shifted away from self-conscious city building toward the management and regulation of increasingly complex city systems of building, transportation, communication, and services. Whereas nineteenth-century city building was often measured by how much it aided commerce by creating settings apart from commerce, twentieth-century planning sought to do so by increasing the efficiency of the city's economic system. This reorientation of planning also increasingly dealt with a plurality of urban constituencies, including the working class. The challenge became less that of building a city and more of managing it. With this change, the importance of the cultivated ideal and the landscapes it created noticeably declined.

Notes

Introduction

1. Carl Sandburg, *Chicago Poems* (New York: Henry Holt, 1916), 3–4. For observations on literary responses to Chicago, see Carl S. Smith, *Chicago and the American Literary Imagination, 1880–1920* (Chicago: University of Chicago Press, 1984), esp. chap. 4, "Business and Art," 57–90.

2. The term applied to United States society is W. E. B. Du Bois's, cited in *Immigrant Women in the Land of Dollars: Life and Culture on the Lower East Side, 1890–1925,* by Elizabeth Ewen (New York: Monthly Review Press, 1985). But also see *Chicago: The History of Its Reputation,* part 1, by Lloyd Lewis, Introduction and part 2, by Henry Justin Smith (New York: Harcourt, Brace, 1929), vi: "the hunt for dollars, women and fame is violent here." Lewis here suggests the term's specific applicability to Chicago.

3. Lewis Mumford, *Sticks and Stones: A Study of American Architecture and Civilization* (New York: Horace Liveright, 1924), 83, 87, 109–10; Mumford's views of American urbanization have been fostered and extended in recent historiography, informed by post–World War II urban crises.

4. Carl Condit, *The Chicago School of Architecture: A History of Commercial and Public Buildings in the Chicago Area, 1875–1925* (Chicago: University of Chicago Press, 1964), 16, says that Mumford characterized 1870s Chicago as "a brutal network of industrial necessities." Condit notes Chicagoans' interest in high culture and their reputation as concerned only with Mammon, 25.

5. See Condit, *Chicago School,* 27; Carl Condit, *The Rise of the Skyscraper* (Chicago: University of Chicago Press, 1952); Joseph Siry, *Carson Pirie Scott: Louis Sullivan and the Chicago Department Store* (Chicago: University of Chicago Press, 1988), 3–12.

6. An identification between Chicago and skyscrapers runs throughout the literature; for example, see Condit, *Chicago School,* 11–12; Mark L. Peisch identifies the "Chicago School" as the "countermovement" to the "Classical Eclecticism" of the World's Fair of 1893, thus suggesting the foreignness of such eclecticism. Peisch, *The Chicago School of Architecture: Early Followers of Sullivan and Wright* (New York: Random House, 1964), 3–4.

7. Students of vernacular architecture have explored the links between buildings and culture. See *Common Places: Readings in American Vernacular Architecture,* ed. Dell Upton and John M. Vlach (Athens: University of Georgia Press, 1986), xiii–xxiv

8. Upton and Vlach refute the idea that the societies that create vernacular architectural forms are necessarily simple. Upton and Vlach, *Common Places,* xv–xvi.

209

9. For an overview of Chicago social history, see Frederic Cople Jaher, *The Urban Establishment: Upper Strata in Boston, New York, Charleston, Chicago, and Los Angeles* (Urbana: University of Illinois Press, 1982), chap. 5; Bessie Louise Pierce, *A History of Chicago,* 3 vols. (New York: Knopf, 1937–56).

10. Jaher, *Urban Establishment,* 453–539; Mark Girouard, *Cities and People: A Social and Architectural History* (New Haven: Yale University Press, 1985), 305–6.

11. Girouard, *Cities and People,* 320–21, notes the reliance of Chicago skyscraper builders on capital from New York City and Boston.

12. Judith Newton, "Family Fortunes: 'New History' and 'New Historicism,'" *Radical History Review* 43 (winter 1989), 8–9, emphasizes that the creation of these spheres constituted the essence of the nineteenth-century's developing middle-class culture. Also on the relation between class formation and a changing gender system, see Christine Stansell, *City of Women: Sex and Class in New York, 1789–1860* (New York: Knopf, 1986), xiii. See also Mary P. Ryan, *Cradle of the Middle Class: The Family in Oneida County, N.Y., 1790–1865* (Cambridge: Cambridge University Press, 1985).

13. Mary P. Ryan, *Women in Public: Between Banners and Ballots, 1825–1880* (Baltimore: Johns Hopkins University Press, 1990); Karen Halttunen, *Confidence Men and Painted Ladies: A Study of Middle-Class Culture in America, 1830–1870* (New Haven: Yale University Press, 1982), 115.

14. Jaher, *Urban Establishment,* 497.

15. Eric Hobsbawm, *The Age of Empire* (New York: Pantheon, 1987), chap. 7.

16. See Helen Lefkowitz Horowitz, *Culture and the City: Cultural Philanthropy in Chicago from the 1880s to 1917* (Lexington: University Press of Kentucky, 1976), 1–48; Jaher, *Urban Establishment,* 463, 500–509.

17. For further discussion, see Daniel M. Bluestone, "Detroit's City Beautiful and the Problem of Commerce," *Journal of the Society of Architectural Historians* 47 (September 1988): 245–62.

Chapter 1
"A Cordon of Verdure"
Gardens, Parks, and Cultivation, 1830–1869

1. Isaac N. Arnold, "William B. Ogden and Early Days in Chicago." Paper Read before the Chicago Historical Society, Tuesday, 20 December 1881 (Chicago: Fergus Printing, 1882); see also Frederic Cople Jaher, *The Urban Establishment* (Urbana: University of Illinois Press, 1982), 454–55; "William B. Ogden," *Biographical Sketches of the Leading Men of Chicago* (Chicago: Wilson & St. Clair, 1868), 11–24.

2. Ibid.

3. Ogden to J. Whitcombs, 18 October 1837, William B. Ogden Papers, Manuscript Division, Chicago Historical Society.

4. Ogden to F. Vanderburgh, 12 December 1840; Ogden to B. H. Mores, 24 January 1838, Ogden Papers.

5. Eighth Census of the United States (1860), Cook County, City of Chicago, Illinois, wards 1 and 2, manuscript copy, roll 164, National Archives, Washington, D.C. Microfilm.

6. Arnold, *Ogden,* 28.

7. Ogden to Mary Haliday, 1 August 1839, Ogden Papers.

8. Arnold, *Ogden;* see also Jaher, *Urban Establishment,* 454–55.

9. Arnold, *Ogden,* 24–26.

10. Eighth Census of the United States (1860).

11. "Chicago Gardening," *Prairie Farmer* 9 (August 1849): 235–37; "Private Gardens in Chicago," *Prairie Farmer* 7 (September-October 1847): 276, 277, 307, 308; "J. Young Scammon," *Chicago Magazine* 1 (March 1857): 43–49; *Chicago Daily Journal,* 3 May 1851.

12. *Weekly Chicago Democrat,* 4 and 18 May 1847.

13. *Daily Chicago Democrat,* 7 March 1849.

14. Richard L. Bushman, "American High-Style and Vernacular Cultures," *Colonial British America: Essays in the New History of the Early Modern Era,* ed. Jack P. Greene and J. R. Pole (Baltimore: Johns Hopkins University Press, 1984), 345–83; Kenneth Cmiel, *Democratic Eloquence: The Fight over Popular Speech in Nineteenth-Century America* (New York: William Morrow, 1990), 37–49, 69–70.

15. Ibid.; and Kevin M. Sweeney, "Mansion People: Kinship, Class, and Architecture in Western Massachusetts in the Mid-Eighteenth Century," *Winterthur Portfolio* 19 (winter 1984): 231–55.

16. Andrew Jackson Downing, *Rural Essays* (New York: G. P. Putnam, 1853), 142; see also *The Artist in American Society: The Formative Years, 1790–1860,* by Neil Harris (New York: George Braziller, 1966), 208–16; Norman Newton, *Design on the Land: The Development of Landscape Architecture* (Cambridge: Harvard University Press, 1971), 260–69.

17. Downing, *Rural Essays,* 110.

18. Downing, *Rural Essays,* 111.

19. "Speech of William B. Ogden, Esq., of Chicago, Delivered before the Board of Commissioners of the Union Pacific Railroad Company at Their Convention, held in Chicago, Illinois, 2 September 1864" (New York: L. H. Bigelow, 1864), 3.

20. *Chicago American,* 1 April 1837.

21. *Chicago American,* 19 September 1835.

22. *Western Citizen,* 3 April 1849; *Gem of the Prairie,* 30 June 1849.

23. *Chicago Daily Journal,* 4 February 1851.

24. Ibid.

25. Resolution printed in *Chicago Democrat,* 4 November 1835; see also *Forever Open, Clear and Free: The Struggle for Chicago's Lakefront,* by Lois Wille (Chicago: Henry Regnery, 1972), 22–23.

26. *Chicago American,* 14 November 1835.

27. Poinsett to Matthew Birchard, 23 April 1839, War Department Records, Letters Sent by Secretary of War relative to Military Affairs File, National Archives, Washington, D.C.

28. Petition signed by 159 people to President Van Buren 29 April 1839, War Department Records, Quarter Master General Reservation Files; Poinsett to Mayor B. W. Raymond, 4 May 1839, War Department Records, Letters Sent by Secretary of War relative to Military Affairs File; Poinsett to Common Council of City of Chicago, 22 May 1839, ibid.

29. *Daily Chicago American,* 1, 7, 13, and 15 June 1839.

30. *Daily Chicago American,* 28 May 1839; Birchard to Poinsett, 25 May 1839, War Department Records, Judge Advocate General Files; Poinsett to Birchard, 14 June 1839, War Department Records, Letters Sent to Secretary of War relative to Military Affairs File.

31. Birchard to Poinsett, 21 June 1839 and 21 November 1840, War Department Records, Judge Advocate General Files.

32. "Samuel H. Kerfoot," in *Leading Men of Chicago,* 313–15; Samuel H. Kerfoot, "History of Lakeview," Chicago Historical Society.

33. *Chicago Daily Journal,* 8 September 1847; *Weekly Chicago Democrat,* 23 March 1847.

34. Bryant to William Ware, 14 September 1835, *The Letters of William Cullen Bryant,* vol. 1, 1809–36, ed. William Cullen Bryant II and Thomas G. Voss (New York: Fordham University Press, 1975), 466–70.

35. "The Battery of Chicago," *Gem of the Prairie,* 30 June 1849; Hans Huth, *Nature and the Americans* (Berkeley: University of California Press, 1957), 54–70; Daniel M. Bluestone, "From Promenade to Park: The Gregarious Origins of Brooklyn's Park Movement," *American Quarterly* 39 (winter 1987): 529–50.

36. *Gem of the Prairie,* 30 June 1849.

37. Bluestone, "From Promenade to Park," 535.

38. "Architecture in the United States," *American Journal of Science and Arts* 17 (January 1830): 105–8.

39. *Daily Chicago Democrat,* 14 January 1851.

40. Downing, *Rural Essays,* 142.

41. *Gem of the Prairie,* 11 August 1849; Bluestone, "From Promenade to Park," 535–38.

42. European visitors particularly noted the fine gradations between social "sets" within the respectable classes in nineteenth-century American cities. Karen Halttunen, *Confidence Men and Painted Women: A Study in Middle-Class Culture in America, 1830–1870* (New Haven: Yale University Press, 1982), 62–63, 96.

43. *Gem of the Prairie,* 30 June 1849; Bluestone, "From Promenade to Park"; see also Halttunen, *Confidence Men and Painted Women,* esp. 93–96, 104–5, 114–16.

44. *Weekly Chicago Democrat,* 16 March 1850.

45. *Daily Chicago Democrat,* 19 July 1853; *Chicago Tribune,* 13 and 26 December 1853, and 11, 16, 25 January, and 8 February 1854.

46. "The Public Parks of Chicago, Their Origin, Former Control and Present Government," *Chicago City Manual,* 1914, 7; Wright is also quoted in "Private Plans for Public Squares: The Origins of Chicago's Park System, 1850–1875," by Glen E. Holt, *Chicago History* 8 (fall 1979): 173. See also Augustine W. Wright, *In Memoriam: John S. Wright* (Chicago: Fergus Printing, 1885), 17; John S. Wright, *Chicago Past, Present, Future: Relations to the Great Interior and to the Continent* (Chicago: Horton & Leonard, 1868), 281.

47. *Chicago Times,* 15 September 1866; see also *Chicago Tribune,* 15 September 1866.

48. Galen Cranz, *The Politics of Park Design: A History of Urban Parks in America* (Cambridge: MIT Press, 1982), esp. 3, 5; Lewis Mumford, *Sticks and Stones: A Study of American Architecture and Civilization* (New York: Horace Liveright, 1924), 95; Ross L. Miller, "The Landscaper's Utopia versus the City: A Mismatch," *New England Quarterly* 49 (June 1976): 179–93; Thomas Bender, *Toward an Urban Vision: Ideas and Institutions in Nineteenth-Century America* (Louisville: University of Kentucky Press, 1975). David Schuyler, *The New Urban Landscape: The Redefinition of City Form in Nineteenth-Century America* (Baltimore: Johns Hopkins University Press, 1986), 59–146, also cites a nostalgic agrarianism but finds that nineteenth-century Americans viewed aspects of urban life as preferable to rural life.

49. Perry Miller, "Nature and the National Ego," *Errand into the Wilderness* (Cambridge: Harvard University Press, 1956), 203–16; Roderick Nash, *Wilderness and the American Mind* (New Haven: Yale University Press, 1982), 44–83; Morton

White and Lucia White, *The Intellectual versus the City* (Cambridge: Harvard University Press, 1962); Bender, *Toward an Urban Vision,* provides a useful historical account of the ways in which Americans attempted to resolve these antithetical notions by invoking images of the middle landscape. He draws his concept of the middle landscape from Leo Marx, *The Machine in the Garden: Technology and the Pastoral Ideal in America* (New York: Oxford University Press, 1964). For an overview of American studies literature on the concept of nature and a useful insight on the *mutability* and flexibility of the nature myths in America, see Barbara Novak, *Nature and Culture: American Landscape and Painting, 1825–1875* (New York: Oxford University Press, l980), 3–17.

50. *Chicago Tribune,* 23 November 1874.

51. Horace Binney Wallace, "Parks and Parking," *Art and Scenery in Europe and Other Papers,* 2d ed. (Philadelphia: J. B. Lippincott, 1868), 315.

52. Second Annual Report of the West Chicago Park Commissioners for the Year Ending 28 February 1871 (Chicago: W. Cravens, 1871), 13.

53. Ogden's letters appear in the *Daily Democratic Press,* 26 January, 20 February, 23 March, and 26 June 1854; Bross's letters appear in the *Chicago Tribune,* 15 November 1867, 24 December 1867, and 1 February 1868.

54. *Chicago Tribune,* 21 October 1859.

55. *The Parks and Property Interests of the City of Chicago* (Chicago: Western News, 1869), 29.

56. H. W. S. Cleveland, *The Public Grounds of Chicago: How to Give Them Character and Expression* (Chicago: Charles D. Lakey, 1869), 3. The 1869 edition of this pamphlet was published without author attribution; later editions carried Cleveland's name.

57. H. W. S. Cleveland, *Landscape Architecture as Applied to the Wants of the West,* ed. Roy Lubove (Pittsburgh: University of Pittsburgh Press, [1873] 1965), 15–17.

58. Olmsted to Alfred T. Field, 11 April 1871, Frederick Law Olmsted Papers, Manuscript Division, Library of Congress, Washington, D.C. All Olmsted material is located in this collection unless otherwise specified.

59. John H. Rauch, *Public Parks: Their Effects upon the Moral, Physical and Sanitary Conditions of the Inhabitants of Large Cities with Special Reference to the City of Chicago* (Chicago: S. C. Griggs, 1869), 31.

60. Victoria Post Ranney, associate editor of the Olmsted Papers, believes that McCagg was the host, based on other letters and materials in the Olmsted collection.

61. Frederick Law Olmsted, "Journey to the West," 1863.

62. *Chicago Tribune,* 23 March 1869.

63. Cleveland to Olmsted, 25 July 1888.

64. Charles S. Spring to Cyrus Hall McCormick, 2 and 24 February 1869, McCormick Papers, Wisconsin Historical Society, Madison.

65. *Chicago Tribune,* 23 March 1869.

66. "J. Young Scammon," *Chicago Magazine* 1 (March 1857): 43–49.

67. *Leading Men of Chicago,* 32.

68. See Scammon, letter to the editor, *Chicago Tribune,* 30 January 1867.

69. *Chicago Times,* 20 February 1867; Bross was a member of the party that accompanied Speaker of the U.S. House of Representatives Schuyler Colfax on a Western tour to consider railroad development. Visiting Yosemite, they heard Olmsted propose establishing it as a park reserve.

70. William Bross to Olmsted, 28 April and 20 July 1866.

71. *Chicago Times,* 7 February 1867.

72. *Chicago Tribune,* 7 and 8 February 1867.

73. As one of Chicago's first park commissioners, Cornell later strongly defended the artistic and design integrity of the Olmsted and Vaux plan for the South Parks; see Cornell to Olmsted, 29 May 1877.

74. *Chicago Times,* 15 September 1866.

75. *Chicago Times,* 17 April 1867.

76. *Chicago Times,* 18 April 1867.

77. *Chicago Times,* 16 April 1867; *Chicago Tribune,* 10 March 1869.

78. George R. Clarke, letter to the editor: *Chicago Tribune,* 10 April 1868.

79. *Chicago Times,* 17 April 1867.

80. *Chicago Tribune,* 10 April 1867.

81. *Chicago Times,* 24 March 1869.

82. Newton, *Design on the Land,* 223–24; George F. Chadwick, *The Park and the Town: Public Landscape in the Nineteenth and Twentieth Centuries* (New York: Praeger, 1966), 33–34, 68, 71, 156. James S. T. Stranahan, president of the Brooklyn Park Commission, noted the influence of these European models. He supported leaving Litchfield Mansion on Prospect Park property, stating that the mansions in Regent's Park "rather added to the beauty of the grounds than detracted from it," *Brooklyn Daily Eagle,* 26 November 1859. In 1869, Boston Alderman Messings proposed a 400-acre buffer area around an 800-acre park to "keep off offensive trades and nuisances," and, giving European precedent a more democratic cast, suggested using the land to "furnish the laboring classes with homes at a low rate," *Boston Daily Advertiser,* 10 November 1869.

83. *Chicago Tribune,* 5 March 1871; Elizabeth Barlow Rogers, *Frederick Law Olmsted's New York* (New York: Praeger, 1972), 13.

84. *Chicago Tribune,* 20 March 1869.

85. *The Parks and Property Interests of the City of Chicago,* 13.

86. Cleveland to Olmsted, 20 March 1885; "Drexel Boulevard—The Great Driveway to the South Parks," *Land Owner* 3 (March 1871): 75–76.

87. Second Annual Report of the West Chicago Park Commissioners, 12.

88. *Chicago Tribune,* 10, 18, and 20 March 1869.

89. Mark Skinner to Mary L. Newberry, 22 February 1871, Blatchford Papers, Newberry Library, Chicago.

90. Gwendolyn Wright, *Moralism and the Model Home: Domestic Architecture and Cultural Conflict in Chicago, 1873–1913* (Chicago: University of Chicago Press, 1980), perceptively discusses the domestic ideal of sanctuary in the nineteenth-century city. See also Margaret Marsh, "From Separation to Togetherness," *Journal of American History* 76 (September 1989): 506–27. People sought in the land economics and regulations encompassed by park development the assurance that "no manufactories or other establishments will ever be erected" and that the neighborhood adjacent to parks would remain "strictly residential"; see "Humboldt Park Residence Property," *Land Owner* 2 (July 1870): 174.

91. *Chicago Tribune,* 19 July 1869; Olmsted, "Journey to the West."

92. *The Parks and Property Interests of the City of Chicago,* 30.

93. Mel Scott, *American City Planning since 1890* (Berkeley: University of California Press, 1969), 1–26; Jon A. Peterson, "Origins of the Comprehensive City Planning Ideal in the United States" (Ph.D. diss., Harvard University, 1967); Thomas S. Hines, *Burnham of Chicago: Architect and Planner* (New York: Oxford University Press, 1974), 313–16; Thomas Adams, *Outline of Town and City Planning* (New York: Russell Sage Foundation, 1935), 167–84.

94. Cynthia Zaitzevsky, *Frederick Law Olmsted and the Boston Park System* (Cambridge: Harvard University Press, 1982), 28, 51.

95. Olmsted, "Journey to the West."

96. "Humboldt Park Residence Property," *Land Owner* 2 (July 1870): 174; "Our Chicago Market Review," *Land Owner* 3 (October 1871): 330–31; "Within and Without the Park Cordon," *Land Owner* 4 (September 1872): 146; *Chicago Tribune,* 6 August 1869 and 20 July 1873.

97. *Chicago Tribune,* 31 December 1866.

98. Cleveland to Olmsted, 7 September 1870; Cleveland, *Landscape Architecture,* 16.

99. Cleveland, *Landscape Architecture,* 26.

100. Cleveland, *Landscape Architecture,* 27.

101. Olmsted, Vaux and Company, Report accompanying Plan for Laying Out the South Park, March 1871 (Chicago: South Park Commission, 1871), 4–5.

102. Second Annual Report of the West Park Commissioners, 12–13; see also Second Report of the Board of Commissioners of the Department of Parks for the City of Boston, 1876 (Boston: Rockwell & Churchill, 1876), 8; Cleveland's view on urban competition with suburbs is in the *Chicago Tribune,* 23 November 1874; for a view of Brooklyn, New York's competition with suburbs, see *Brooklyn Daily Eagle,* 17 November 1859; on park-suburb continuity, see Edward K. Spann, *The New Metropolis: New York, 1840–1857* (New York: Columbia University Press, 1981), 172–73, 176–79; Lewis Mumford, *The City in History* (New York: Harcourt, Brace, World, 1961), 489–93.

103. John H. Rauch to Olmsted, 21 and 22 April 1869; *Chicago Tribune,* 29 April 1872. Suburban Cicero's opposition to West Parks is noted in the *Chicago Tribune,* 17 March 1869.

104. *Chicago Tribune,* 26 October 1867.

105. John H. Rauch to Olmsted, 10, 12, and 20 January 1869.

106. *Chicago Tribune,* 18 February 1869, 23 November 1874.

107. *Chicago Tribune,* 14 and 16 March 1869

108. Annual Report of the Commissioners of Lincoln Park, 1 April 1879 (Chicago: Geo. E. Marshall, 1879), 9.

109. Scott, *American City Planning,* 10–12; see also Cranz, *Politics of Park Design,* 61–99.

110. *Brooklyn Daily Eagle,* 28 November 1859; Cleveland, Public Parks, Radial Avenues, and Boulevards: Outline Plan of a Park System for the City of St. Paul, Comprised in Two Addresses Delivered before the Common Council and Chamber of Commerce, 24 June 1872 and 19 June 1885 (St. Paul: Globe Job Office, 1885), 7.

111. Chicago Board of Public Works, Annual Reports, 13th, 31 March 1874, 20; 14th, 31 March 1875, 14–20; 15th, 31 December 1875, 21 (Chicago: Thompson, 1874–76).

112. *Chicago Tribune,* 15 January and 4 December 1868.

113. *Chicago Tribune,* 28 March 1873.

114. *Chicago Tribune,* 18 and 24 February 1869.

115. On the limits to respectable women's mobility, see Halttunen, *Confidence Men,* 115; Mary P. Ryan, *Women in Public: Between Banners and Ballots, 1825–1880* (Baltimore: Johns Hopkins University Press, 1990), 58–94.

116. Cranz, *Politics of Park Design,* 183–84.

117. Paul Boyer, *Urban Masses and Moral Order in America, 1820–1920* (Cambridge: Harvard University Press, 1978), esp. chap 18, which considers the City Beautiful movement as positive environmentalism.

118. Bryant to William Ware, 14 September 1835, *The Letters of William Cullen Bryant,* 466–70.

119. *Chicago Tribune,* 24 March 1869.

120. Roy Rosenzweig, *Eight Hours for What We Will: Workers and Leisure in an Industrial City, 1870–1920* (Cambridge: Cambridge University Press, 1983), 127–90; Boyer, *Urban Masses and Moral Order,* 242–51; Benjamin McArthur, "The Chicago Playground Movement: A Neglected Feature of Social Justice," *Social Service Review* (September 1975): 376–95.

Chapter 2
"A Different Style of Beauty"
Park Designs, 1865–1880

1. Olmsted, Vaux and Company, Report accompanying Plan for Laying Out the South Park, March 1871 (Chicago: South Park Commission, 1871), 14.

2. H. W. S. Cleveland, *The Public Grounds of Chicago: How to Give Them Character and Expression* (Chicago: Charles D. Lakey, 1869), 13.

3. Olmsted, Vaux and Company, Preliminary Report upon the Proposed Suburban Village of Riverside, near Chicago (New York: Sutton, Browne, 1868), 3.

4. John Munn Journal, 30 April 1849, Manuscript Division, Chicago Historical Society.

5. Cleveland, *Public Grounds of Chicago,* 13.

6. *Chicago Tribune,* 14 March 1869.

7. *Chicago Tribune,* 9, 13, and 17 May 1869; Vaux to Olmsted, 5 and 10 October 1869; James P. Root to Olmsted, Vaux and Company, 5 April 1870; Olmsted Papers, Manuscript Division, Library of Congress, Washington, D.C. All Olmsted material is located in this collection unless otherwise specified.

8. Olmsted and Vaux, Report for South Park, 9.

9. Cynthia Zaitzevsky, *Frederick Law Olmsted and the Boston Park System* (Cambridge: Harvard University Press, 1982), 73–75.

10. Charles E. Beveridge, "Frederick Law Olmsted's Theory of Landscape Design," *Nineteenth Century* 3 (summer 1977): 38–45; see also Bruce Kelly, "Art of the Olmsted Landscape," in *Art of the Olmsted Landscape,* by Bruce Kelly, Gail Travis Guillet, and Mary Ellen W. Hern (New York: New York City Landmarks Preservation Commission and Art Publisher, 1981), 4–71.

11. Olmsted and Vaux, Report for South Park, 9.

12. Beveridge, "Olmsted's Theory of Landscape Design," 38–43.

13. Olmsted and Vaux, Report for South Park, 7.

14. Sixth Annual Report of the Board of Commissioners of Prospect Park, Brooklyn, January 1866 (Brooklyn: I. Van Anden, 1866), 16; David Schuyler, *The New Urban Landscape: The Redefinition of City Form in Nineteenth-Century America* (Baltimore: Johns Hopkins University Press, 1986), 119–25.

15. Olmsted and Vaux, Description of a Plan for the Improvement of the Central Park: 'Greensward' 1858, reprinted in *Landscape into Cityscape: Frederick Law Olmsted's Plans for a Greater New York City,* ed. Albert Fein (Ithaca, N.Y.: Cornell University Press, 1967), 73–75.

16. Andrew Jackson Downing, *Rural Essays* (New York: G. P. Putnam), 10.

17. Olmsted and Vaux, Riverside, 16–17.

18. The suggestion for a boat connection to the parks was initially made in *The Parks and Property Interests of the City of Chicago* (Chicago: Western News, 1869), 36; Olmsted and Vaux, Report for South Park, 14, 15, 27, 33.

19. Report of the South Park Commission, 1 December 1876 to 1 December 1877 (n.d., n.p.), 5–7; Minutes of the South Park Commission, Manuscript Division, Chicago Historical Society, esp. 21 July 1877; L. B. Sidway, "Early History of the South Parks," Report of the South Park Commission, 1 March 1908 to 28 February 1909 (n.p., n.d.), 60–70.

20. H. W. S. Cleveland, "Landscape Architects' Report," Report of the South Park Commission to the Board of County Commissioners of Cook County, 1 December 1872 to 1 December 1873 (n.p., n.d.), 12–20.

21. Hale to Olmsted, 12 March and May 1894; Ellsworth to Olmsted, 25 April 1894; *Chicago Times,* 9 March 1894; *Chicago Herald,* 7 March 1894; *Inter-Ocean,* 14 March 1894.

22. *Chicago Tribune,* 2 March and 8 June 1872.

23. *Chicago Tribune,* 8 June 1872.

24. For a description of Washington Park, see the *Chicago Tribune,* 17 December 1876.

25. Cleveland to Olmsted, 10 September 1872.

26. Clarence Pullen, "The Parks and Parkways of Chicago," *Harpers' Weekly* 35 (30 May 1891): 413, 416; Cleveland to Olmsted, 25 July 1888; for a defense of Kanst's designs, see Andreas Simon, *Chicago, the Garden City: Its Magnificent Parks, Boulevards and Cemeteries* (Chicago: Franz Gindele, 1895), 10.

27. Minutes of the South Park Commission, 21 July and 9 November 1877, Manuscript Division, Chicago Historical Society.

28. Schuyler, *The New Urban Landscape,* 37–56.

29. *Weekly Chicago Democrat,* 22 March 1848.

30. Thomas Bender, *Toward an Urban Vision: Ideas and Institutions in Nineteenth-Century America* (Louisville: University of Kentucky Press, 1975), 81–88; Hans Huth, *Nature and the Americans* (Berkeley: University of California Press, 1957), 66–69; Neil Harris, *The Artist in American Society: The Formative Years, 1790–1860* (New York: George Braziller, 1966), 200–208; Downing, *Rural Essays,* 77, 81, 155; Schuyler, *New Urban Landscape,* 37–56; *Chicago Tribune,* 29 July and 15 October 1859, 15 April 1861. Landscape architect William Saunders, associated with the improvements of Philadelphia's Laurel Hill Cemetery, provided the plan for Rose Hill after being recommended to the Rose Hill trustees by the president of Laurel Hill Cemetery, *Chicago Tribune,* 30 and 31 August 1860; *Charter of Graceland Cemetery* (Chicago: James Barnet, 1861).

31. Board of Public Works, Annual Report, 2d, for Year Ending April 1st, 1863 (Chicago: Tribune Book and Job Printing Office, 1863), 23, 70; also Annual Report, 4th, 1865, 29; Lincoln Park Common Council Proceedings, 1864, Manuscript, Chicago Historical Society.

32. Board of Public Works, Annual Reports, 5–9, 1866–70.

33. At first the city refused to turn the park land over to the Lincoln Park Commission; some residents challenged the assessment accounts for the land purchase and improvement. The constitutionality of the acts was also challenged. See *Chicago Tribune,* 17 September 1869, 29 April 1872.

34. McCagg to Olmsted, 1 March 1869.

35. McCagg failed in his efforts to have Olmsted design Lincoln Park. Both Jenney and Cleveland attempted to take over improvement of the park from Nelson, but the Lincoln Park Commission continued to employ Nelson as the designer and contractor for the work. French to Olmsted, 17 October 1874.

36. I. J. Bryan, Report of the Commissioners and a History of Lincoln Park (Chicago, 1899); *Guide to Lincoln Park: A Handbook for the Use of Visitors with Descriptions of Objects of Interest* (Chicago: King & Kelly, 1889).

37. *Chicago Tribune,* 16 July 1869.

38. *Chicago Tribune,* 22 and 26 August 1869.

39. The Wolcott and Fox Report is reprinted in *Chicago Tribune,* 16 July 1869.

40. *Chicago Tribune,* 5 November 1869.

41. Jenney to Olmsted, 2 December 1865.

42. Sanford E. Loring and W. L. B. Jenney, *Principles and Practice of Architecture, Comprising Forty-Six Folio Plates of Plans, Elevations and Details of Churches, Dwellings and Stores Constructed by the Authors. Also an Explanation and Illustrations of the French System of Apartment Houses, and Dwellings for the Laboring Classes, together with Copious Text* (Chicago: Cobb, Pritchard, 1869), preface, 14.

43. Jenney corresponded with Olmsted on the designs for the West Parks. See Jenney to Olmsted, 24 January 1870.

44. "Report of Jenney, Schermerhorn and Bogart, Architects and Engineers," in Second Annual Report of West Chicago Park Commission for Year Ending 28 February 1871 (Chicago: W. Cravens, 1871), 51–81.

45. Ibid.; see also Theodore Turak, *William Le Baron Jenney: A Pioneer of Modern Architecture* (Ann Arbor: UMI Research Publications, 1986), 75–112.

46. Cleveland considered the park system a "curious fact." *Chicago Tribune,* 23 November 1874.

47. Glen E. Holt, "Private Plans for Public Squares: The Origins of Chicago's Park System, 1850–1875," *Chicago History* 8 (fall 1979): 173; John S. Wright, *Chicago Past, Present, Future: Relations to the Great Interior and to the Continent* (Chicago: Leonard & Horton, 1868), 281. For a discussion of American parkway systems, see Schuyler, *New Urban Landscape,* 126–46.

48. *Chicago Tribune,* 16 June 1872.

49. *Chicago Tribune,* 23 April 1871.

50. *Chicago Tribune,* 6 and 13 April 1873.

51. Cleveland to Olmsted, 18 and 30 June 1874.

52. *Chicago Tribune,* 23 November 1874.

53. "S. J. Walker, Esq. and What He Is Doing for Chicago," *Land Owner* 4 (June 1872): 90. Walker hired Swain Nelson to supervise planting on Ashland Avenue. *Chicago Tribune,* 4 May and 27 July 1873.

54. H. W. S. Cleveland, "Parks and Boulevards in Cities," *Lakeside Monthly* 7 (May 1872): 412–14; *Chicago Tribune,* 26 October 1867, 23 November 1874.

55. H. W. S. Cleveland, *Outline Plan of a Park System for the City of St. Paul* (St. Paul: H. M. Smyth, 1885), 25–26.

56. *Chicago Tribune,* 20 July 1873.

57. Cleveland to Olmsted, 8 April 1869.

58. Ibid.; and *Chicago Tribune,* 23 November 1874; Cleveland, *Public Grounds of Chicago,* 15–17, 19.

59. When designers such as Cleveland, Jenney, and Olmsted turned their attention to boulevard development, they turned their attention to Europe. Cleveland drew heavily upon William Robinson's book *The Parks, Promenades, and Gardens of Paris* (London: John Murray, 1869), as well as to the related work and designs of Haussmann's landscape architect, Aldolphe Alphand, who in the late 1860s published the monumental *Les Promenades de Paris.* Cleveland particularly shared Robinson's enthusiasm for smaller parks and gardens and landscaped boulevards and tree-lined streets, which introduced refreshing scenes of nature into the city center. Cleveland frequently quoted Robinson's view that nature in Paris "follows the street builders with trees, turns the little square into gardens unsurpassed for good taste and beauty, drops down graceful fountains here and there, and margins them with flowers": see Robinson, 2, 36, 112, 116, 117. See also Cleveland's view in the *Chicago Tribune,* 10 March 1873. Frederick Law Olmsted, *New American Cyclopedia,* 1861, s.v. "Park," ed. George Ripley and Charles A. Dana (New York: D. Appleton, 1861), vol. 12, 768–75, reveals Olmsted's attentiveness to European boulevard and park design and his popularization of these forms. See also Cleveland to Olmsted, 18 and 30 June 1874.

60. Olmsted, "Park."

61. Quotations from Olmsted, Vaux and Company, Preliminary Report respecting a Public Park in Buffalo, and a Copy of the Act of the Legislature Authorizing Its Establishment (Buffalo: Matthew and Warren, 1869), 25–27. For a discussion of formal boulevard design, see Sixth Annual Report of the Board of Commissioners of Prospect Park, Brooklyn, January 1866 (Brooklyn: I. Van Anden, 1866), 33, 37; Francis R. Kowsky, "Municipal Parks and City Planning: Frederick Law Olmsted's Buffalo Park and Parkway System," *Journal of the Society of Architectural Historians* 46 (March 1987): 49–64.

62. The quotation is from Olmsted, "Park"; Olmsted, "Public Parks and the Enlargement of Towns," *Journal of Social Science* 2 (1871): 1–36, reprint of an address read before the American Social Science Association, Boston, 25 February 1870; Olmsted and Vaux, "Riverside," 12.

63. Olmsted and Vaux, "Riverside," 12.

64. Olmsted to Riverside Improvement Company, 27 April 1869; "Riverside Boulevard," *Land Owner* 1 (September 1869): 60. The dimensions given are those found on the plan in this article.

65. These gathering spots echoed Olmsted's earlier descriptions of the drives along Florence's Arno River, where carriages could gather at the promenade hours to hear music performed; Olmsted, "Park."

66. Cornell to Calvert Vaux, 5 October 1869; Cornell to Olmsted, 10 January 1877, 12 March 1878, and 6 October 1879. Departures from the Olmsted and Vaux plan led to the removal of two lines of trees from both planted areas and destruction of the bridle paths and walkways. The commission later returned to the initial plan, including the return of the three lines of trees on either side of the carriageway.

67. Olmsted and Vaux, Report for South Park, 21.

68. Simon, *Chicago, the Garden City,* 56; Everett Chamberlin, *Chicago and Its Suburbs* (Chicago: T. A. Hungerford, 1874), 320.

69. Cleveland to Olmsted, 11 July 1874.

70. *Chicago Tribune,* 17 July 1873.

71. "The Chicago of the 'Societist,'" *Lakeside Monthly* 10 (October 1873): 330; *Chicago Tribune,* 16 June 1872, 17 December 1876.

72. *Boston Daily Advertiser,* supplement, 4 December 1869; Simon, *Chicago, the Garden City,* 48.

73. South Park Commission, Annual Report, 1 December 1875 to 1 December 1876, 34.

74. Minutes of the South Park Commission, 23 July and 23 August 1873, Manuscript Division, Chicago Historical Society; *Chicago Tribune,* 17 July 1873.

75. *Chicago Tribune,* 16 June 1872.

76. Cleveland, quoted in the *Chicago Tribune,* 23 November 1874; see also Olmsted, "Public Parks and the Enlargement of Towns."

77. Galen Cranz, *The Politics of Park Design: A History of Urban Parks in America* (Cambridge: MIT Press, 1982), 13–15, 19–24.

78. Minutes of the South Park Commission, 23 July 1873.

79. Simon, *Chicago, the Garden City,* 60; see "Introductory," *Guide to Lincoln Park.*

80. *Chicago Tribune,* 6 July 1873.

81. Cleveland to Olmsted, 23 June 1889.

82. Cleveland to Olmsted, 25 July 1888.

Chapter 3
"A Parallel Moral Power"
Churches, 1830–1895

1. *Chicago Tribune,* 11 August 1865.

2. *Interior* 1 (24 March 1870): 8.

3. For a discussion of the social, cultural, and geographical centrality of the Anglican parish in colonial Virginia, see Dell Upton, *Holy Things and Profane: Anglican Parish Churches in Colonial Virginia* (Cambridge: MIT Press, 1986); see also Carl Feiss, "Early American Public Squares," *Town and Square from the Agora to the Village Green,* ed. Paul Zucker (New York: Columbia University Press, 1959), 237–55; Anthony N. B. Garvan, *Architecture and Town Planning in Colonial Connecticut* (New Haven: Yale University Press, 1951), 42, 51, 61–66, 130; John R. Stilgoe, "The Puritan Townscape: Ideal and Reality," *Landscape* 20 (spring 1976): 3–7; William Butler, "Another City upon a Hill: Litchfield, Connecticut, and the Colonial Revival," *The Colonial Revival in America,* ed. Alan Axelrod (New York: W. W. Norton, 1985), 13–51; John W. Reps, *Town Planning in Frontier America* (Princeton: Princeton University Press, 1965), 103–14.

4. Reps, *Town Planning,* 240–72.

5. Sydney E. Ahlstrom, *A Religious History of the American People* (New Haven: Yale University Press, 1972), explores the Protestant tendency to embrace plainness; Ahlstrom notes the Puritans' aversion to towers, 148; Garvan, *Architecture and Town Planning,* 130–40, 144–45. William N. Hosley, Jr., "Architecture," *The Great River: Art and Society of the Connecticut River Valley, 1635–1820,* ed. Gerald W. R. Ward and William N. Hosley, Jr. (Hartford, Conn.: Wadsworth Atheneum, 1985), 66, notes the revival of churchlike features in meeting houses; see also Kevin M. Sweeney, "Mansion People: Kinship, Class, and Architecture in Western Massachusetts in the Mid-Eighteenth Century," *Winterthur Portfolio* 19 (winter 1984): 231–55; J. Frederick Kelly, *Early Connecticut Meeting Houses,* 2 vols. (New York: Columbia University Press, 1948).

6. St. James Parish Vestry Minutes. Reverend Hallam is quoted in the typescript of Rev. Wm. H. Vibbert, "Historical Discourse," 26 October 1884, Archives of the Episcopal Diocese of Chicago.

7. *The Second Presbyterian Church of Chicago, 1 June 1842 to 1 June 1892* (Chicago: Knight, Leonard, 1892), 28–30; Second Presbyterian Church, Minutes of the Session and Board of Trustees, 1842–67, Second Presbyterian Church, Chicago.

8. Vibbert, "Historical Discourse."

9. Alfred T. Andreas, *History of Chicago, From the Earliest Period to the Present Time,* 3 vols. (Chicago: A. T. Andreas, 1884–86), 1: 334–36.

10. Lewis Cass Aldrich and Frank R. Holmes, eds., *History of Windsor County, Vermont* (Syracuse, N.Y.: D. Mason, 1891), 324–25.

11. Robert A. Hallam, *Annals of St. James's Church, New London, for One Hundred and Fifty Years* (Hartford, Conn.: Church Press, 1873), 70.

12. *Chicago Tribune,* 22 September 1872.

13. "John M. Van Osdel," *Biographical Sketches of the Leading Men of Chicago* (Chicago: Wilson & St. Clair, 1868), 91–95.

14. Van Osdel contributed to the pattern book genre with his *Carpenter's Own Book* (Baltimore: John W. Woods, 1834).

15. *Daily Democrat,* 28 January 1848.

16. *Weekly Chicago Democrat,* 8 May 1849.

17. *Convention of Ministers and Delegates of the Congregational Church of the United States: A Book of Plans for Churches and Parsonages, Published under the Direction of the Central Committee, Appointed by the General Conference, October, 1853* (New York: D. Burgess, 1853).

18. *Weekly Chicago Democrat,* 28 January 1848.

19. *Daily Democrat,* 7 August 1849.

20. *Weekly Chicago Democrat,* 28 January 1848.

21. *Daily Democrat,* 7 August 1849.

22. John Moses and Joseph Kirkland, eds., *History of Chicago, Illinois,* 2 vols. (Chicago: Munsell, 1895), vol. 1, 624–28; John Barber and Henry Howe, *Historical Collections of the State of New York* (New York: S. Tuttle, 1841).

23. Eliphalet Wickes Blatchford, "Memorial Sketch of John Chandler Williams and His Wife Mary M. Williams," n.d. [c. 1865], Chicago Historical Society; "William H. Brown," *Chicago Magazine* (March 1857), quoted in William H. Bushnell, *Biographical Sketches of Some of the Early Settlers of the City of Chicago* (Chicago: Fergus, 1876), 6–10; Henry Hall, *The History of Auburn* (Auburn: Dennis Bros., 1869); Thomas B. Carter, "Some Facts and Incidents in the Early Life of Thomas Butler Carter from Boyhood On, Till Now, But Especially during the Past Fifty years from the Date of Arrival in Chicago 15 September 1838," 1888, Chicago Historical Society; Samuel L. Tuttle, *A History of the Presbyterian Church, Madison, N.J.: A Discourse Delivered on Thanksgiving Day, 23 November 1854* (New York: M. W. Dodd, 1855), 19, 57–67.

24. Quoted in Gilbert J. Garragham, *The Catholic Church in Chicago* (Chicago: Loyola University Press, 1921), 127.

25. *Catholic Almanac for 1849* (Baltimore: F. Lucas, Jr., 1848), 136; Garragham, *Catholic Church in Chicago,* 113–19.

26. *Metropolitan Catholic Almanac and Laity's Directory,* 1845 (Baltimore: Fielding Lucas, 1844), 113.

27. *Fifth Annual Review of the Commerce, Manufactures, and the Public and Private Improvements of Chicago [for the Year 1856]* (Chicago: Democratic Press, 1857), 8.

28. George S. Phillips, *Chicago and Her Churches* (Chicago: E. B. Myers and Chandler, 1868), 395.

29. Rev. Z. M. Humphrey, "Historical Sketch" [1867], in Philo Adams Otis, *The First Presbyterian Church, 1833–1913* (Chicago: Fleming H. Revell, 1913), 35.

30. *Chicago Daily Democrat,* 8 May 1857.

31. *History of the First Baptist Church, Chicago* (Chicago: R. R. Donnelley & Son, 1889), 34–35.

32. *Chicago Tribune,* 22 September 1872.

33. Minutes of the Second Presbyterian Church Trustees, 28 April 1868, Second Presbyterian Church, Chicago, Illinois.

34. *Manual of the Second Presbyterian Church, Chicago* (Chicago: Charles Scott, 1880), 8. The lot was later traded for one on Michigan and 20th Street where the congregation moved in 1872.

35. Eighth Census of the United States (1860), Cook County, City of Chicago, Illinois, wards 1 and 2. Manuscript copy, roll 164, National Archives, Washington, D.C. Microfilm.

36. *Fifth Annual Review of Commerce,* 11.

37. *Chicago Tribune,* 15 December 1873.

38. Frederick Law Olmsted, "Journey to the West," 1863, Frederick Law Olmsted Papers, Library of Congress, Washington, D.C.

39. The classic article on domesticity is by Barbara Welter, "The Cult of True Womanhood, 1820–1860," *American Quarterly* 18 (summer 1966), 131–75; Nancy F. Cott, *The Bonds of Womanhood: "Woman's Sphere" in New England, 1780–1835* (New Haven: Yale University Press, 1977), chap. 2; Linda Kerber, "Separate Spheres, Female Worlds, Woman's Place: The Rhetoric of Women's History," *Journal of American History* 75 (June 1988): 9–39.

40. Ann Douglas, *The Feminization of American Culture* (New York: Knopf, 1977).

41. *Interior* 5 (8 January 1874): 8.

42. Mary Ryan, *Women in Public: Between Banners and Ballots, 1825–1880* (Baltimore: Johns Hopkins University Press, 1990), 58–94; Karen Halttunen, *Confidence Men and Painted Women: A Study of Middle-Class Culture in America, 1830–1870* (New Haven: Yale University Press, 1982), 115.

43. Halttunen, *Confidence Men and Painted Ladies,* 58–59;

44. *Interior* 23 (4 February 1892): 8–9.

45. Phillips, *Chicago and Her Churches,* 191. See also the *Chicago Tribune,* 7 February 1867, 7 September 1873.

46. "The Churches of Chicago," *Land Owner* 5 (September 1873): 169; also quoted in *Wing's Illustrated Travellers' and Visitors' Handbook to the City of Chicago* (Chicago: J. M. Wing, 1874).

47. *Chicago Tribune,* 20 August 1866.

48. *Journal of the Twenty-First Annual Convention of the Diocese of Illinois* (1858), 37.

49. *Interior* 5 (8 January 1874): 8.

50. See Margaret Marsh, "From Separation to Togetherness," *Journal of American History* 76 (September 1989): 506–27 on the difference between men's and women's attitudes toward the downtown.

51. Phoebe B. Stanton, *Pugin* (New York: Viking Press, 1972); idem, *The Gothic Revival and American Church Architecture: An Episode in Taste, 1840–1856* (Baltimore: Johns Hopkins University Press, 1968); Basil Clarke, *Church Builders of the Nineteenth Century* (New York: Macmillan, 1938); Kenneth Clark, *The Gothic Revival* (New York: Harper and Row, [1928] 1974); A. Welby Pugin, *Contrasts: Or, A parallel between the Noble Edifices of the Middle Ages, and the Corresponding Buildings of the Present Day, Showing the Present Decay of Taste* (New York: Humanities Press, [1841] 1969).

52. Frederick Clarke Withers, *Church Architecture* (New York: A. J. Bicknell, 1873), vii. See Francis R. Kowsky, *The Architecture of Frederick Clarke Withers, and the Progress of the Gothic Revival in America after 1850* (Middletown: Wesleyan University Press, 1980).

53. Andrew Jackson Downing, *Rural Essays* (New York: G. P. Putnam, 1853), 245–46.

54. William H. Pierson, Jr., *American Buildings and Their Architects: Technology and the Picturesque, the Corporate and Early Gothic Styles* (Garden City: Doubleday, 1978), 153.

55. Ibid.; and Robert Maccigrosso, *American Gothic: The Mind and Art of Ralph Adams Cram* (Washington, D.C.: University Press of America, 1980).

56. Neil Harris, *The Artist in American Society: The Formative Years, 1790–1860* (New York: George Braziller, 1966); Roger B. Stein, *John Ruskin and Aesthetic Thought in America, 1840–1900* (Cambridge: Harvard University Press, 1967).

57. Rev. R. W. Patterson, *The Place Where God Records His Name: A Discourse Delivered at the Dedication of the Second Presbyterian Church on Friday Evening, 24 January 1851, Chicago, Illinois* (Chicago: S. C. Griggs, 1851), 21.

58. *Chicago Tribune,* 28 July 1872.

59. *Chicago Tribune,* 27 March 1869.

60. "William W. Boyington," *Biographical Sketches of the Leading Men of Chicago,* 215–22. The best sources for graphic views of Chicago's churches are: Andreas, *History of Chicago;* Paul Thomas Gilbert and Charles Lee Bryson, *Chicago and Its Makers* (Chicago: F. Mendelsohn, 1929); *Chicago Illustrated* (Chicago: Jevne & Almini, 1867–68).

61. *Chicago Illustrated; Chicago Tribune,* 18 May 1869.

62. *Chicago Tribune,* 20 June 1869.

63. David Swing, "The Chicago of the Christian," *Lakeside Monthly* 10 (October 1873): 337.

64. *Chicago Tribune,* 16 March 1866.

65. *Chicago Tribune,* 17 January 1873.

66. *Chicago Tribune,* 11 November 1871.

67. *Chicago Tribune,* 17 January 1873.

68. On liberal religion in the nineteenth century, see Ahlstrom, *Religious History,* chap. 46.

69. Phillips, *Chicago and Her Churches,* 318–22.

70. David T. Van Zanten, "Jacob Wrey Mould: Echoes of Owen Jones and the High Victorian Styles in New York, 1853–1865," *Journal of the Society of Architectural Historians* 27 (March 1969): 41–57.

71. Jacob Wrey Mould to Bishop Henry John Whitehouse, 24 January 1853; Whitehouse to Mould, 4 March 1853. Unless otherwise cited, Whitehouse material is located in the Archives of the Episcopal Diocese of Chicago.

72. Mould to Whitehouse, 24 January 1853.

73. Clarkson to Whitehouse, 6 January 1857; Strong to Whitehouse, 20 September 1860.

74. In the 1850s, Whitehouse himself had considered taking up residence in one of the Michigan Avenue row houses: see Clarkson to Whitehouse, 6 January 1857.

75. Phillips, *Chicago and Her Churches,* 318–22; Andreas, *History of Chicago,* 1: 407–9.

76. "Tearing Down Churches," *Diocese of Chicago* 1 (October 1885): 1.

77. Ibid.; and the *Chicago Times,* quoted in *Diocese of Chicago* 1 (November 1885): 1; see also the *Chicago Tribune,* 20 September 1896.

78. Records of Grace Episcopal Church Vestry, 21 October 1867, Grace Episcopal Church, Chicago.

79. "Tearing Down Churches," 1.

80. *Journal of the Thirty-Second Annual Convention of the Diocese of Illinois* (1869), 14.

81. *Journal of the Thirty-Sixth Annual Convention of the Diocese of Illinois* (1873), 47.

82. *Journal of the Thirtieth Annual Convention of the Diocese of Illinois* (1867), 85.

83. "History of Trinity Church Chicago," *Diocese of Chicago* 4 (May 1888): 2; *Chicago Tribune,* 18 May 1869.

84. *Chicago Tribune,* 19 September 1869.

85. "The Heart of the City," *Standard,* 1 April 1880.

86. For this rhetorical distinction, see Patterson, *Where God Records His Name,* 11–12.

87. *Chicago Tribune,* 4 January 1869; and see also 5 June 1870.

88. *Chicago Tribune,* 26 November 1877.

89. Ahlstrom, *Religious History,* 325, 437–38, notes that American Protestantism also included forms of worship such as field preaching and camp meetings, which showed a disregard for church edifices.

90. *Chicago Tribune,* 26 February 1876.

91. H. F. Parker, "Elegant Churches," *Advance,* 29 October 1868.

92. "A Million Dollar Church," *Chicago Alliance,* 29 May 1875; *Chicago Tribune,* 5 September 1875.

93. *Chicago Tribune,* 8 July and 7 October 1877.

94. *Chicago Tribune,* 3 September 1877; and *Standard,* 17 September 1874, 11 March 1889.

95. *Chicago Tribune,* 26 November 1877.

96. Rev. H. N. Powers to St. John's Warden and Vestrymen, 7 September 1877, St. John's Episcopal Parish Records, Archives of the Episcopal Diocese of Chicago.

97. "The Church Critic," *Chicago Pulpit* 2 (20 July 1872): 44.

98. *Standard,* 24 February 1876.

99. William Wallace Everts, *The House of God: Or Claims of Public Worship* (New York: American Tract Society, 1872), 60–61.

100. *Chicago Tribune,* 10 February 1873; *Inter-Ocean,* 10 February 1873; *Living Church* 8 (9 May 1885): 70.

101. *Chicago Tribune,* 13 November 1871.

102. *Chicago Tribune,* 5 August 1858.

103. *Chicago Tribune,* 13 July 1868.

104. *Chicago Tribune,* 29 April 1859.

105. Andreas, *Chicago,* 2: 425; *Chicago Tribune,* 5 August 1858, 29 April 1859.

106. First Methodist Episcopal Church, *Seventy-Fifth Anniversary: Diamond Jubilee Souvenir Program* (Chicago, 1910).

107. *Chicago Tribune,* 5 August 1858, 29 April 1859.

108. *Interior* 23 (21 July 1892): 26.

109. George Lane, *Chicago Churches and Synagogues* (Chicago: Loyola University Press, 1981), 163.

110. *Chicago Tribune,* 3 August 1873; *Standard,* 15 July 1875; Lafayette Wallace Case, *History of the North Star Mission, North Star Baptist Church, and the La Salle Avenue Baptist Church* (Chicago: Lenington, 1897), 6–8.

111. Minutes of the Second Presbyterian Church Trustees, 1 February 1869, 7 February 1870, Second Presbyterian Church, Chicago; *Chicago Tribune,* 25 April 1871.

112. "Endowments," *Diocese of Chicago* 8 (October 1892): 3.

113. William R. Hutchison, *The Modernist Impulse in American Protestantism* (Cambridge: Harvard University Press, 1976).

114. Phillips, *Chicago and Her Churches*, 333–42.

115. Joseph Fort Newton, *David Swing: Poet Preacher* (Chicago: Unity Publishing, 1909); William R. Hutchison, "Disapproval of Chicago: The Symbolic Trial of David Swing," *Journal of American History* 59 (June 1972): 30–47.

116. *Chicago Tribune,* 8 December 1873.

117. *Chicago Tribune,* 23 February 1886.

118. See the sermon in Newton, *Swing,* 157; David Swing, "Devotion and Work," in *David Swing: A Memorial Volume,* ed. Helen Swing Starring (Chicago: F. Tennyson Neely, 1894), 147–66.

119. *Chicago Tribune,* 22 February 1886; Newton, *Swing,* 116–18, 207.

120. [David Swing,] "Limestone Christianity," *Chicago Alliance,* 4 April 1874.

121. *Chicago Tribune,* 14 December 1875.

122. *Detroit Post,* quoted in the *Chicago Tribune,* 5 December 1875.

123. *Chicago Tribune,* 6 December 1875.

124. *Chicago Tribune,* 5 December 1880.

125. Quoted in *In Memory of George Benedict Carpenter, Died 7 January 1881,* Chicago Historical Society, 1881.

126. Austin Bierbower, *Life and Sermons of Dr. H. W. Thomas, including the Discourses on which He Is Charged with Heresy* (Chicago: Smith & Forbes, 1880); Thomas Wakefield Goodspeed, "Hiram Washington Thomas," *University of Chicago Biographical Sketches* (Chicago: University of Chicago Press, 1922), 335–58.

127. *Chicago Tribune,* 8 November 1880.

128. *Chicago Tribune,* 17 October 1881.

129. Records of the First Presbyterian Church, 1 December 1895, 1 December 1901, 1 December 1903, and 1 June 1912, First Presbyterian Church, Chicago.

130. *Interior* 23 (28 April 1892): 19.

Chapter 4
"A City under One Roof"
Skyscrapers, 1880–1895

1. John J. Flinn, *Chicago, The Marvelous City of the West: A History, an Encyclopedia, and a Guide* (Chicago: Standard Guide, 1892), 570.

2. "The Chamber of Commerce Building," *Economist* 3 (8 March 1890): 266.

3. Lawrence Wakefield, *All Our Yesterdays* (Traverse City: Village Press, 1977).

4. See Lawrence Wakefield and Lucille Wakefield, *Historic Traverse City Houses* (Midland: McKay Press, 1978).

5. The quotation is in the Minutes of the National Women's Christian Temperance Union, 1887, cclxxiv; for a discussion of the skyscraper's dominance of American urban form, see Spiro Kostof, "The Skyscraper City," *Design Quarterly* 140 (1988): 33–35.

6. Robert Bruegmann, "The Myth of the 'Chicago School,'" *Thresholds.* Forthcoming. See also Thomas A. P. van Leeuwen, *The Skyward Trend of Thought* (Cambridge: MIT Press, 1988), 20–29.

7. Sigfried Giedion, *Space, Time and Architecture: The Growth of a New Tradition* (Cambridge: Harvard University Press, 1941); Carl W. Condit, *The Rise of the Skyscraper* (Chicago: University of Chicago Press, 1952).

8. Condit, *Skyscraper,* 2. The most recent elaboration of this concept comes in Ross Miller's *American Apocalypse: The Great Fire and the Myth of Chicago* (Chicago: University of Chicago Press, 1990), 149–60. Miller argues that American nineteenth-century architects had "two insecure identities": that of the professional and that of the romantic, inventive designer; he concludes, "By incorporating style, American architects might have temporarily solved their identity crises, but there was something absurd about it all. Images that comforted the architect bore no relation to the job at hand of permanently rebuilding a city like Chicago" (p. 160).

9. Giedion, *Space, Time and Architecture,* 24.

10. Ibid.

11. Giedion, *Space, Time and Architecture,* 303–4.

12. Condit, *Skyscraper,* 16.

13. United States Census Office, *The Statistics of the Population of the United States, 9th Census, 1870* (Washington, D.C.: GPO, 1872), vol. 1; idem, *Statistics of the Population of the United States at the Tenth Census* (Washington, D.C.: GPO, 1883), 870, table 36; Department of Commerce and Labor, Bureau of the Census, *Special Reports, Occupations at the Twelfth Census* (Washington, D.C.: GPO, 1904).

14. Alfred D. Chandler, Jr., *The Visible Hand: The Managerial Revolution in American Business* (Cambridge: Harvard University Press, 1977); Olivier Zunz, *Making America Corporate, 1870–1920* (Chicago: University of Chicago Press, 1990), 103–48.

15. *Chicago Tribune,* 28 October 1888; see also Perry R. Duis, *Chicago Creating New Traditions* (Chicago: Chicago Historical Society, 1976), 21, 29.

16. In the 1880s and 1890s, buildings of ten stories were considered skyscrapers; see *Call Board Bulletin,* 11 October 1888; *Industrial Chicago,* 4 vols. (Chicago: Goodspeed Publishing, 1891), vol. 1, 168. The term *commercial style,* used interchangeably with *skyscraper,* is said to apply to buildings of seven stories and over in *Morris' Dictionary of Chicago and Vicinity* (Chicago: Frank M. Morris, 1891), 28–29.

17. Gerald R. Larson and Roula Mouroudellis Geraniotis, "Toward a Better Understanding of the Evolution of the Iron Skeleton Frame in Chicago," *Journal of the Society of Architectural Historians* 46 (March 1987): 40–41; Rosemarie Haag Bletter, "The Invention of the Skyscraper: Notes on Its Diverse Histories," *Assemblage* 2 (February 1987): 110–11. For a somewhat different view on this subject, see, Cesar Pelli, "Skyscrapers," *Perspecta* 18 (1982): 134–51; Robert A. M. Stern, Gregory Gilmartin, and John Montague Massengale, *New York 1900* (New York: Rizzoli, 1983), 147–52, 455*n*19.

18. *Chicago Tribune,* 16 February 1890, quoted in *The Architecture of John Wellborn Root,* by Donald Hoffmann (Baltimore: Johns Hopkins University Press, 1973), 85.

19. Morris, *Dictionary of Chicago,* 29.

20. Ogden to George S. Boutwell, 27 January 1872, Public Building Service Records, record group 121, entry 26, box 8, National Archives, Washington, D.C.

21. William S. McCormick to Cyrus Hall McCormick, 4 October 1863. This letter and all other McCormick business and real estate materials are, unless otherwise specified, located in the McCormick Papers, Manuscript Division, State Historical Society of Wisconsin, Madison.

22. McCormick Company to C. Mohan, 15 March 1872.

23. W. J. Hanna to H. O. Goodrich, 21 March 1873.

24. McCormick Company to A. W. Nichols, 23 January 1883.

25. Charles A. Spring, Jr., to Mrs. Cyrus H. McCormick, 31 October 1879.

26. Cyrus Hall McCormick, Jr., Diary, 26 July 1879.

27. *Chicago Tribune,* 9 March 1884.

28. *Chicago Office Building Directory* (Chicago: Chicago Office Building Directory Company, 1892).

29. John J. Flinn, *The Standard Guide of Chicago* (Chicago: Standard Guide Company, 1893), 47.

30. Flinn, *Standard Guide,* 55.

31. Newspaper descriptions are the single most important source of information on nineteenth-century public views of the skyscraper. There are both feature articles and notices of building plans in the real estate sections of the papers. Other important sources are building rental brochures. The Chicago Historical Society has the best collection of these, including brochures for the Board of Trade, Masonic Temple Continental, Woman's Temple, Venetian, Counselman, Isabella, Columbus Memorial, and Chicago, Burlington, and Quincy Railroad buildings.

32. John W. Root, "A Great Architectural Problem," *Inland Architect* 15 (June 1890): 71.

33. William H. Jordy, *American Buildings and Their Architects: Progressive and Academic Ideals at the Turn of the Twentieth Century* (Garden City, N.Y.: Doubleday, 1972), 73.

34. *Chicago Tribune,* 28 October 1888.

35. Flinn, *Chicago,* 1892, 574.

36. Flinn, *Standard Guide,* 1893, 197.

37. "The Chamber of Commerce Building," 266; Flinn, *Chicago,* 1892, 570–71.

38. See *Collection of Photographs of "Ornamental Iron" Executed by the Winslow Bros. Co., Chicago* (Chicago: Winslow Bros., 1893).

39. John F. Kasson, "The Aesthetics of Machinery," in his *Civilizing the Machine* (New York: Grossman Publishers, 1976): 137–80.

40. Charles S. Spring, Jr., to Mrs. Cyrus Hall McCormick, 13 November 1879.

41. "Mosaic Floors," *Inland Architect* 17 (July 1891): 71.

42. *Industrial Chicago,* vol. 1, 197.

43. [Van H.] Higgins and [Henry J.] Furber, *The Columbus Memorial Building cor. State and Washington Streets, Chicago* (Chicago: Higgins and Furber, 1893). Pamphlet.

44. *Chicago Tribune,* 23 September 1891.

45. [Emory A. Storrs,] "Judge Van H. Higgins," in *The Bench and Bar of Chicago* (Chicago: American Biographical Publishing, 1883), 580–84.

46. "Henry J. Furber," *Bench and Bar of Chicago,* 510.

47. *Chicago Tribune,* 23 September 1891.

48. "The Phenix Insurance Company Building," *Inland Architect* 10 (September 1887): supplement.

49. *Chicago Tribune,* 18 June 1890; "The Masonic Temple," *Economist* 3 (21 June 1890): 807.

50. Eric J. Hobsbawm, *The Age of Empire, 1875–1914* (New York: Pantheon, 1987), 172.

51. *Chicago Tribune,* 24 April 1892.

52. *Chicago Tribune,* 24 April 1892, 1 May 1892, and 8 May 1892.

53. For a recent review of the debate, see Larson and Geraniotis, "The Iron Skeleton Frame in Chicago"; Condit, *Skyscraper,* 114; Jordy, *American Buildings,* 20–22; Theodore Turak, *William Le Baron Jenney: A Pioneer of Modern Architecture* (Ann Arbor: UMI Research Publications, 1986), 237–63.

54. "Removal of the Railway Age Offices," *Railway Age* 11 (8 April 1886): 185–86.

55. *Chicago Tribune,* 30 October 1892.

56. McCormick Real Estate Ledgers, 1863–70.

57. Alfred T. Andreas, *History of Chicago,* 3 vols. (Chicago: A. T. Andreas, 1884–86), vol. 3, 64.

58. "The Fowler, Goodell, Walters Block," *Landowner* 4 (March 1872): 38, 41.

59. "The Republic Life Insurance Company," *Landowner* 2 (June 1870): 144; "The Republic Life and John V. Farwell's Government Buildings," *Landowner* 4 (April 1872): 54–55.

60. *Chicago Tribune,* 26 November 1882.

61. *The Hayes Skylight and Blind Company [Catalog]* (New York: Livingston Middledith, 1887), 5; *Chicago Tribune,* 1 May 1881.

62. "Open Board of Trade," *Inland Architect* 3 (June 1884): supplement; see also *Hayes Skylight and Blind Company,* 6.

63. *Chicago Tribune,* 15 July 1888.

64. Flinn, *Chicago,* 1892, 570.

65. Rand McNally, *Bird's-Eye View and Guide to Chicago* (Chicago, 1893), 136.

66. *Chicago Tribune,* 28 October 1888.

67. *Chicago Office Building Directory,* 1892), 79.

68. *Chicago Tribune,* 28 October 1888.

69. "The Phenix Insurance Company Building."

70. Barbara Guttman Rosenkrantz, *Public Health and the State* (Cambridge: Harvard University Press, 1972), 9, 10, 25, 75–77; George Rosen, *A History of Public Health* (New York: M. D. Publications, 1958), 294–95.

71. "Chicago," *American Architect and Building News* 34 (21 November 1891): 118.

72. Morris, *Dictionary of Chicago,* 131.

73. *Chicago Tribune,* 22 May and 19 June 1892; J. Lincoln Steffens, "The Modern Business Building," *Scribner's Magazine* 22 (July 1987): 37–66.

74. "List of Employees under Charge of Real Estate Committee, 23 March 1886," Chicago Board of Trade Papers, Manuscript Archives, University of Illinois, Chicago Circle.

75. Zunz, *Making America Corporate,* 103–48.

76. "The New Pullman Office and Apartment Building," *Western Manufacturer* 12 (31 March 1884): 41.

77. "The Pullman Palace-Car Company," *National Car-Builder* 4 (February 1873): 38.

78. Julian Ralph, "The Highest of All Roof Gardens," *Harper's Weekly Magazine* 36 (3 September 1892): 855.

79. Flinn, *Chicago,* 1892, 571.

80. *Chicago Tribune,* 16 October 1892.

81. *Chicago Tribune,* 7 November 1886.

82. *Industrial Chicago,* vol. 1, 168; on stylistic diversity, see Julian Ralph, "Chicago—The Main Exhibit," *Harper's Monthly Magazine* 84 (February 1892): 425–36, and Franklin H. Head, "The Heart of Chicago," *New England Magazine,* n.s., 6 (July 1892): 561.

83. Louis Sullivan, "The Tall Office Building Artistically Considered," *Lippincott's Magazine* 57 (March 1896): 403.

84. Sullivan, "Tall Office Building," 403.

85. Sullivan, "Tall Office Building," 405.

86. Mark Girouard, *Cities and People: A Social and Architectural History* (New Haven: Yale University Press, 1985), 322, argues that Chicago skyscrapers do not adequately represent the era's preference for ornament. Chicago was conceived as "a branch-office city," so that New York-based companies invested less heavily in office buildings there than in New York City. By this account, the plainer "Chicago style" resulted from a lack of capital rather than from aesthetic preference. This argument, however, requires more systematic documentation of both comparative corporate enterprise and of the relative costs of ornamented and plain styles.

87. William Le Baron Jenney to Olmsted, 2 December 1865, Frederick Law Olmsted Papers, Manuscript Division, Library of Congress, Washington, D.C.

88. William Le Baron Jenney and Sanford E. Loring, *Principles and Practice of Architecture, comprising Forty-six Folio Plates of Plans, Elevations and Details of Churches, Dwellings and Stores, Constructed by the Authors* (Chicago: Cobb, Pritchard, 1869), 9, 14. See also Turak, *William Le Baron Jenney.*

89. Root, "A Great Architectural Problem," 67.

90. Root, "A Great Architectural Problem," 67–71.

91. Ralph, "Chicago—The Main Exhibit," 425.

92. Jordy, *American Buildings,* 71–75.

93. Ralph, "Chicago—The Main Exhibit"; Head, "Heart of Chicago," 561.

94. Edward K. Spann, *The New Metropolis: New York, 1840–1857* (New York: Columbia University Press, 1981), 98.

95. See Joseph Siry, *Carson Pirie Scott: Louis Sullivan and the Chicago Department Store* (Chicago: University of Chicago Press, 1988); Russell Lewis, "Everything under One Roof: World's Fairs and Department Stores in Paris and Chicago," *Chicago History* 12 (fall 1983): 28–47; William R. Leach, "Transformations in a Culture of Consumption: Women and Department Stores, 1890–1925," *Journal of American History* 71 (September 1984): 319–42; Susan Porter Benson, *Counter Cultures: Saleswomen, Managers, and Customers in American Department Stores* (Urbana: University of Illinois Press, 1986).

96. Benson, *Counter Cultures,* treats the department store as a workplace for saleswomen, managers, and others. See Michael B. Miller, *The Bon Marche: Bourgeois Culture and the Department Store, 1869–1920* (Princeton: Princeton University Press, 1981) on the new sets of social relations, social roles, and cultural perceptions involved in the rise of the department store.

97. For example, see "A City under One Roof—The Masonic Temple," *Scientific American* 70 (10 February 1894): 81–82.

98. Flinn, *Chicago,* 1892, 571.

99. Ralph, "Chicago—The Main Exhibit," 425–36.

100. *Call Board Bulletin,* 11 October 1888.

101. *Industrial Chicago,* vol. 1, 170.

102. Sigmund Krausz, *Street Types of Chicago* (Chicago: Max Stern, 1892), v; see also Perry Duis, "Whose City? Public and Private Places in Nineteenth-Century Chicago," *Chicago History* 12 (spring 1983): 2–27.

103. Flinn, *Chicago,* 1892, 571; see also Daniel Bluestone "'The Pushcart Evil': Peddlers, Merchants and New York City's Streets, 1890–1940," *Journal of Urban History* (November 1991).

104. Hoffmann, *John Wellborn Root,* 196.

105. *Economist* 3 (21 June 1890): 807.

106. Charles Moore, *Daniel H. Burnham: Architect, Planner of Cities,* 2 vols. (Boston: Houghton Mifflin, 1921); Thomas S. Hines, *Burnham of Chicago: Architect and Planner* (New York: Oxford University Press, 1974).

107. John Colman Adams, "What a Great City Might Be—Lessons from the White City," *New England Magazine,* n.s., 14 (March 1896): 3–13; Charles Zeublin, "The White City and After," *Chautauquan* 38 (December 1903): 373–84; Reid Badger, *The Great American Fair: The World's Columbian Exposition and American Culture* (Chicago: N. Hall, 1979).

108. "The New Pullman Office Building," *Chicago Journal of Commerce* 44 (14 March 1884): 65–77.

109. "New Building of the Pullman Palace Car Company," *Railway Age* 9 (15 May 1884): 318; "The New Pullman Office and Apartment Building," *Western Manufacturer* 12 (31 March 1884): 1.

110. "Sky Dwellings," *American Builder and Journal of Art* 3 (April 1871): 380.

111. *Chicago Office Building Directory,* 92.

112. Minutes of the National Women's Christian Temperance Union (1887), cclxxiv.

113. *Daily Railway Bulletin,* 15 March 1886.

114. Higgins and Furber, *The Columbus Memorial Building.*

115. *Chicago Tribune:* 9 June 1889; 4, 11, 13, 15, and 18 October 1891; 5, 11, and 18 November 1891; 18 December 1891; 17 February 1892.

116. M. A. Lane, "High Buildings in Chicago," *Harper's Weekly Magazine* 35 (31 October 1891): 853–56; Ernest Flagg, "The Dangers of High Buildings," *Cosmopolitan* 21 (May 1896): 70–79; "Chicago," *American Architect and Building News* 25 (22 June 1889): 293; "High Buildings Again," *Economist* 6 (10 October 1891): 610; Ray Stannard Baker, "The Modern Skyscraper," *Munsey's Magazine* 22 (October 1899): 48–58; "Deliberations of the Architects," *Economist* 6 (21 November 1891): 857–58; "A Halt Called on the Skyscraper," *Scribner's Magazine* 19 (March 1896): 395–96; "The High Building Peril," *American Architect and Building News* 13 (19 May 1883): 237; Harry W. Bringhurst, "Fire and the Modern Skyscraper," *Scientific American* 80 (21 January 1899): 39; "Our Tall Buildings," *Sanitary Engineer* 15 (15 March 1887): 131; R. Kerr, "On the Lofty Buildings of New York City," *Sanitary Engineer* 9 (3 and 10 January 1884): 113, 114, 137, 138; "The Art Critic and the Tall Building," *Scientific American* 80 (28 January 1899): 50; Owen Brainard, "The Modern Tall Building," *Chautauquan* 26 (November 1897): 132–39.

117. *Moran's Dictionary of Chicago and Its Vicinity* (Chicago: George E. Moran, 1892), 48.

Chapter 5
"Less of Pork and More of Culture": Civic and Cultural Chicago, 1850–1905

1. Montgomery Schuyler, "A Critique of the Works of Adler and Sullivan, D. H. Burnham and Company, Henry Ives Cobb," *Architectural Record, Great American Architects Series No. 2* (February 1896): 3–4; see also Julian Ralph, *Harper's Chicago and the World's Fair* (New York: Harper & Brothers, 1893), 1–4.

2. Daniel H. Burnham and Edward H. Bennett, *Plan of Chicago* (Chicago: Commercial Club, 1909), 4. For a recent historical account echoing Burnham, see Mark Girouard, *Cities and People: A Social and Architectural History* (New Haven: Yale University Press, 1985), 353–55.

3. William H. Wilson, *The City Beautiful Movement* (Baltimore: Johns Hopkins University Press, 1989); William H. Wilson, *The City Beautiful Movement in Kansas City* (Columbia: University of Missouri Pres, 1964); Jon A. Peterson, "The City Beautiful Movement: Forgotten Origins and Lost Meanings," *Journal of Urban History* 2 (August 1976): 415–34; Richard E. Foglesong, *Planning the Capitalist City: The Colonial Era to the 1920s* (Princeton: Princeton University Press, 1986), 124–66; William H. Wilson, "The Ideology, Aesthetic, and Politics of the City Beautiful Movement," *The Rise of Modern Urban Planning, 1880–1914,* ed. Anthony Sutcliffe (New York: St. Martin's Press, 1980), 71–98.

4. *Daily Democrat,* 1 January and 12 March 1851; *Daily Journal,* 4 February 1851.

5. John W. Reps, *The Making of Urban America: A History of City Planning in the United States* (Princeton: Princeton University Press, 1965).

6. *Daily Journal,* 7 December 1850.

7. *Daily Journal,* 11 December 1850.

8. *Daily Democrat,* 2 June 1851.

9. *Daily Democrat,* 30 January 1851; *Daily Journal,* 22 May 1851.

10. *Chicago Tribune,* 6 November 1868.

11. *Chicago Tribune,* 11, 13 December 1867 and 14 March 1868.

12. *Chicago Tribune,* 14 January 1868.

13. *Chicago Tribune,* 11 February 1868.

14. *Chicago Tribune,* 25 February 1868.

15. *Chicago Tribune,* 6 and 9 November 1867.

16. "Our Street Architecture: Public Buildings II," *American Builder and Journal of Art* 1 (December 1868): 13–14.

17. James Guthrie to William D. Snowhook, 19 August 1854, Public Buildings Service Records, General Correspondence, record group 121, entry 26, box 48, National Archives, Washington, D.C.

18. Report of the Post Master General and the Secretary of the Interior to the President, 19 December 1854, quoted in Chicago Petition to James Guthrie, 10 February 1855, Public Buildings Service Records, General Correspondence, record group 121, entry 26, box 48,

19. Proposal no. 4, Alexander White, Chicago Customhouse, Public Buildings Service, General Correspondence, record group 121, entry 26, box 48.

20. [Citizens' Petition Against Site Purchase,] c. January, 1855, Common Council Resolution, 18 December 1854; [Citizens' Petition Supporting Site Purchase,] 10 February 1855, Public Buildings Service Records, General Correspondence, record group 121, entry 26, box 48. See also *Chicago Tribune,* 19 January 1855.

21. Isaac Cook to James Guthrie, 26 May 1855, "Illinois Chicago, Customhouse (Old)," Public Buildings Service Records, Title Papers, record group 121.

22. Alexander H. Bowman to William Bigler, 2 March 1858, Public Buildings Service Records, record group 121, entry 6, 2:494–96.

23. See drawings in Files of the Supervising Architect, Public Buildings Service Records, record group 121, entry 35; see also Daniel Bluestone, "Civic and Aesthetic Reserve: Ammi B. Young's 1850s Federal Customhouse Designs," *Winterthur Portfolio* 25 (summer/autumn 1990): 131–56.

24. Benjamin Lombard to Howell Cobb, 26 October 1860; Lombard to Luther Haven, 20 April 1864; S. M. Clark to Ammi B. Young, 15 October 1860; Public Buildings Service Records, record group 121, entry 26, box 48; Jevne & Almini, *Chicago Illustrated* (Chicago, 1867–68).

25. Benjamin Lombard to Howell Cobb, 26 October 1860; Benjamin Lombard to Luther Haven, 20 April 1864.

26. S. M. Clark to Ammi B. Young, 15 October 1860.

27. Agreement affirmed 5 August 1872, *Chicago Tribune,* 27 July 1872.

28. "The Coming City Hall," *Landowner* 5 (November 1873): 203.

29. "Urbs in Horto," *Designs and Plans for Court House and City Hall, to Be Erected in Chicago* (Chicago: Knight and Leonard, 1873); Thomas E. Tallmadge, *Architecture in Old Chicago* (Chicago: University of Chicago Press, 1941), 125–30.

30. *Chicago Tribune,* 5 July 1877.

31. Editorial, *American Builder* 12 (March 1876): 53.

32. *Chicago Tribune,* 15 December 1879, 27 August 1875.

33. Joseph Medill to George S. Boutwell, 30 December 1871; James E. McLean to Medill, 30 December 1871; Public Buildings Service Records, General Correspondence, record group 121, entry 26, box 48.

34. William Sooy Smith, George B. Post, and Orlando W. Norcross to B. H. Bristow, 15 June 1875, Public Buildings Service Records, General Correspondence, record group 121, entry 26, box 48.

35. Gurdon P. Randall to George B. Boutwell, 19 June 1875, Public Buildings Service Records, General Correspondence, record group 121, entry 26, box 48.

36. See, for example, Ernest Flagg, "Public Buildings," in *Proceedings of the Third National Conference on City Planning* (Boston: The University Press, 1911), 42–52.

37. Thomas Hoyne to George S. Boutwell, 1 April 1872, Public Buildings Service Records, General Correspondence, record group 121, entry 26, box 48.

38. *Chicago Tribune,* 20 March 1881; *Inter Ocean,* 17 March 1881.

39. *Chicago Tribune,* 27 March and 4 May 1881.

40. *Chicago Tribune,* 6, 12, and 13 March 1881.

41. *Chicago Tribune,* 27 March 1881.

42. Ibid.

43. Ibid.

44. *Chicago Tribune,* 11 August 1865.

45. Quoted in Isaac E. Adams, *Life of Emery A. Storrs* (Philadelphia: Hubbard Brothers, 1886), 606–7, reprint of article from *New York Graphic* (March 1879).

46. *Chicago Tribune,* 27 March 1881.

47. Ibid.

48. Ibid.

49. Ibid., *Chicago Tribune,* 10 April and 10, 11 October 1881.

50. "Information for Architects," in *Proceedings of the Chicago Public Library from 12 July 1890 to 25 June 1892* (Chicago, 1892), 191–92; *Chicago Tribune,* 27 March 1881.

51. William H. Jordy, *American Buildings and Their Architects: Progressive and Academic Ideals at the Turn of the Twentieth Century* (Garden City, N.Y.: Doubleday, 1872), 356–62.

52. "Information for Architects," 191.

53. "The Chicago Public Library," *Inland Architect* 18 (November 1891): 20; "The Competition for the Chicago Public Library," *Inland Architect* 19 (March 1892): 24–25.

54. *Chicago Tribune,* 5 February 1892; *Chicago Journal,* 5 February 1892.

55. United States Department of the Interior, Bureau of Education, *Circulars of Information of the Bureau of Education,* no. 1 (1881); [William F. Poole,] *The Construction of Library Buildings* (Washington: GPO, 1881), 14–15; William F. Poole, "Points of Agreement in Library Architecture: Discussion," *Library Journal* 16 (December 1891): 97–120; idem, *Remarks on Library Construction* (Chicago: Jansen, McClurg, 1884).

56. Henry I. Sheldon to E. W. Blatchford, 7 October 1887, reprinted in *Proceedings of the Trustees of the Newberry Library, for the Year Ending 5 January 1890* (Chicago: Knight and Leonard, 1890), 3.

57. *Inter Ocean,* 7 February 1892.

58. *Chicago Tribune,* 7 February 1892; see also 6 February 1892.

59. *Chicago Tribune,* 7 February 1892.

60. *Chicago Tribune,* 27 March 1881.

61. "Chicago Public Library," *Harper's Weekly* 41 (18 September 1897): 924, 934.

62. Nineteenth Annual Report of the Board of Directors of the Chicago Public Library, June 1891 (Chicago, 1891), 8.

63. Art Institute of Chicago, *Information about the Art Institute of Chicago* (Chicago, 1900).

64. Art Institute of Chicago, Annual Report of the Trustees, for the Year Ending 5 June 1888 (Chicago: John Morris Company, 1888), 8.

65. *Chicago American,* 17 July 1890.

66. *Inter Ocean,* 20 November 1887.

67. *Chicago Tribune,* 29 April 1887.

68. Art Institute of Chicago, Annual Report of the Trustees, for the Year Ending 4 June 1889 (Chicago: S. A. Maxwell, 1889).

69. Art Institute of Chicago, Annual Report of the Trustees, for the Year Ending June 2, 1891 (Chicago: Knight and Leonard, 1892).

70. John J. Flinn, *The Standard Guide of Chicago* (Chicago: Standard Guide Company, 1893), 130, 166.

71. *Chicago Tribune,* 26 August and 14 September 1894.

72. *Chicago Tribune,* 14 September 1894.

73. Ibid.

74. *Chicago Tribune,* 16 September 1894.

75. *Chicago Tribune,* 20 September 1894.

76. *Chicago Tribune,* 16 September 1894.

77. *Chicago Herald,* 28 September 1890; *Chicago News,* 30 September 1890.

78. *Chicago Tribune,* 25 January 1895.

79. *Chicago Tribune,* 8 and 19 January 1892.

80. L. J. Barr to Washington Hesing, 5 January 1895, Public Buildings Service Records, General Correspondence, record group 121, entry 26, box 48.

81. *Chicago Tribune,* 25 January 1895.

82. Ibid.

83. *Chicago Tribune,* 13 September 1896.

84. *Chicago Tribune,* 12 October 1896; *Inter Ocean,* 12 October 1896.

85. *Chicago Tribune,* 12 October 1896.

Chapter 6
"The Keystone of the Arch"
The Civic Lakefront, 1893–1909

1. *Chicago Tribune,* 23 August 1896.

2. Ernest Flagg, "Public Buildings," in *Proceedings of the Third National Conference on City Planning* (Boston: The University Press, 1911), 49. See Mardges Bacon, *Ernest Flagg: Beaux-Art Architect and Urban Reformer* (Cambridge: MIT Press, 1986), 209–33.

3. Daniel H. Burnham and Edward H. Bennett, *Plan of Chicago* (Chicago: Commercial Club, 1909), 19. See also David H. Pinkney, *Napoleon III and the Rebuilding of Paris* (Princeton: Princeton University Press, 1958); Howard Saalman, *Haussmann: Paris Transformed* (New York: George Braziller, 1971); Carl E. Schorske, "The Ringstrasse, Its Critics, and the Birth of Urban Modernism," in *Fin-De Siecle Vienna: Politics and Culture* (New York: Knopf, 1980), 24–115; Donald J. Olsen, *The City as a Work of Art* (New Haven: Yale University Press, 1985), 44–45, 69–79.

4. Reid Badger, *The Great American Fair: The World's Columbian Exposition and American Culture* (Chicago: Nelson Hall, 1979).

5. *Chicago Tribune,* 23 September 1894.

6. *Inter Ocean,* 10 August 1895.

7. As an architect, Normand S. Patton exhibited a talent for conceptualizing the possibilities of the civic landscape. In 1874, Patton graduated from the Massachusetts Institute of Technology in architecture and moved to Chicago to establish an architectural practice. Two handsome, Romanesque-style library buildings, the Scoville Institute (1885) in Oak Park, Illinois, and the Hackley Public Library (1888) in Muskegon, Michigan, were the earliest of Patton's more than fifty library designs, many of which were executed in connection with Andrew Carnegie's library philanthropy. In the late 1890s, Patton worked as the chief architect for the Chicago public schools, designing numerous monumental classical-style buildings. His practice also included campus plans and buildings for several American colleges. Civic buildings and their settings filled Patton's thoughts and work; see "Patton and Miller," Architects File, Burnham Library, Art Institute of Chicago.

8. Normand S. Patton to Olmsted, Olmsted and Eliot, 17 September 1895, Olmsted Associates Papers, Manuscript Division, Library of Congress, Washington, D.C.

9. "Association Notes," *Inland Architect* 24 (October 1894): 28.

10. *Chicago Tribune,* 27 October 1894.

11. Edgar Weston Brent, *Martin B. Madden: Public Servant* (Chicago, 1901), 175–85; *Chicago Tribune,* 16 and 29 December 1894.

12. *Chicago Tribune,* 30 November 1894; Brent, *Madden,* 183.

13. A transcript of the meeting was published in the *Chicago Tribune,* 30 December 1894.

14. *Chicago Tribune,* 4 June 1895; see also 3 October 1895 issue.

15. One of the earliest presentations of this history is by Charles Moore, *Daniel H. Burnham: Architect Planner of Cities,* 2 vols. (Boston: Houghton Mifflin, 1921), vol. 2, 98–99; Thomas S. Hines, *Burnham of Chicago: Architect and Planner* (New York: Oxford University Press, 1974), 4, 138, 313–16; Norman Newton, *Design on the Land: The Development of Landscape Architecture* (Cambridge: Harvard University Press, 1971), 416; Mel Scott, *American City Planning Since 1890* (Berkeley: University of California Press, 1969), 37–38; Joan E. Draper, *Edward H. Bennett: Architect and City Planner, 1874–1954* (Chicago: Art Institute of Chicago, 1982), 14; Carl W. Condit, *Chicago, 1910–1929: Building, Planning, and Urban Technology* (Chicago: University of Chicago Press, 1973), 59–63.

16. Edward H. Bennett to John Rothwell Slater, 27 July 1907, Edward H. Bennett Papers, Burnham Library, Art Institute of Chicago; John Rothwell Slater, "Making a City into a Metropolis," *World To-Day* 13 (September 1907): 884–92.

17. "Too Hasty Action by County Board," *Inland Architect* 25 (October 1894): 21.

18. *Chicago Tribune,* 30 December 1894.

19. See William H. Wilson, *The City Beautiful Movement* (Baltimore: Johns Hopkins University Press, 1989), 9–34.

20. *Inter Ocean,* 10 August 1895.

21. Ibid.; see also "Chicago Architectural Club," *Inland Architect* 27 (February 1896): 8. The 1896 Architectural Club competition was for a design of a music pavilion for the lakefront.

22. Daniel H. Burnham, [Review of Work,] Auditorium Banquet Hall Address, 25 January 1908, Edward H. Bennett Papers.

23. *Chicago Tribune,* 4 June 1895.

24. Peter B. Wight to John Charles Olmsted, 8 August 1903, Olmsted Associates Papers; see also Patton's objection, *Chicago Tribune,* 9 August 1896.

25. Paul Boyer, *Urban Masses and Moral Order in America, 1820–1920* (Cambridge: Harvard University Press, 1978), 261–76.

26. *Chicago Tribune,* 30 December 1894.

27. *Inter Ocean,* 11 October 1896.

28. *Chicago Tribune,* 28 November 1894.

29. Paul Avrich, *The Haymarket Tragedy* (Princeton: Princeton University Press, 1984).

30. William T. Stead, *If Christ Came to Chicago: A Plea for the Union of All Who Love in the Service of All Who Suffer* (London: Review of Reviews, 1894).

31. *Chicago Tribune,* 13 November 1893.

32. Perry Duis, *The Saloon: Public Drinking in Chicago and Boston, 1880–1920* (Urbana: University of Illinois Press, 1983).

33. Boyer, *Urban Masses*, 261–83.

34. *Ceremonies at the Unveiling of the Statue of Abraham Lincoln, at Lincoln Park, Chicago, Illinois, 22 October 1887* (Chicago: Brown, Pettibone, n.d.), 10–11.

35. *Chicago Tribune*, 10 and 31 January 1886.

36. *Chicago Tribune*, 4 and 8 October 1891; *Inter Ocean*, 7 and 8 October 1891.

37. *Inter Ocean*, 8 October 1891.

38. Ibid.

39. This conclusion arises from numerous statements of dedication-day speakers; see, for example, *Chicago Tribune*, 26 November 1893, 27 September 1896.

40. James L. Riedy, *Chicago Sculpture* (Urbana: University of Illinois Press, 1981), 23–40; see also *Public Sculpture and the Civic Ideal in New York City, 1890–1930*, by Michele H. Bogart (Chicago: University of Chicago Press, 1989), 40–47.

41. *Chicago Tribune*, 4 June 1895.

42. *Dedication of the Ferguson Fountain of the Great Lakes, Chicago, 9 September 1913* (Chicago, 1913), 30–31.

43. *Chicago Tribune*, 16 and 19 April 1905.

44. Riedy, *Chicago Sculpture*, 9.

45. Burnham and Bennett, *Plan of Chicago*, 100.

46. Burnham and Bennett, *Plan of Chicago*, 117.

47. A selection of these earlier studies includes: Bion J. Arnold, *Report on the Engineering and Operating Features of the Chicago Transportation Problem* (New York: McGraw Publishing, 1905); Frederic A. Delano, *Chicago Railway Terminals: A Suggested Solution for the Chicago Terminal Problem* (Chicago, 1904); *Report of the Special Park Commission to the City Council of Chicago on the Subject of a Metropolitan Park System*, comp. Dwight Heald Perkins (Chicago: W. J. Hartman, 1905); *Chicago Tribune*, 6 and 7 January 1905, reports the study of the "Boulevard Link Plan," a plan that later occupied a central position in the 1909 Plan of Chicago.

48. "Proceeding of the 201st Regular Meeting of the Commercial Club of Chicago, 25 January 1908," Edward H. Bennett Papers; Slater, "Making a City into A Metropolis," 889.

49. Burnham and Bennett, *Plan of Chicago*, 117.

50. Ibid.

51. Burnham and Bennett, *Plan of Chicago*, 116.

52. Charles Mulford Robinson, *Modern Civic Art or the City Made Beautiful* (New York: G. P. Putnam's Sons, 1903), 87, 132; see also Guy Kirkham, "The Importance and Value of Civic Centers," *The Grouping of Public Buildings, Municipal Art Society of Hartford, Connecticut*, bulletin no. 2, comp. Frederick L. Ford (Hartford: Municipal Art Society, 1904), 49.

53. Burnham and Bennett, *Plan of Chicago*, 18, 20, 21.

54. Jon A. Peterson, "The Nation's First Comprehensive City Plan: A Political Analysis of the McMillan Plan for Washington, D.C., 1900–1902," *American Planning Association Journal* 51 (1985): 134–50; Hines, *Burnham of Chicago*, 125–216.

55. Robinson, *Modern Civic Art*, 82.

56. Robinson, *Modern Civic Art*, 87, 132; see also Kirkham, "The Importance and Value of Civic Centers," 49.

57. Bennett, "Public and Quasi-Public Buildings," in *City Planning: A Series of Papers presenting the Essential Elements of a City Plan*, ed. John Nolen (New York: D. Appleton, 1916), 110, 114.

58. Daniel H. Burnham and Edward H. Bennett, "Notes on Improvement and Adornment of San Francisco: The Civic Center," c. 1905, Edward H. Bennett Papers, manuscript notes.

59. Daniel H. Burnham, "The Commercial Value of Beauty," *Architects and Builders Journal* 3 (March 1902): 20.

60. Burnham to George Mason, 6 June 1908, Burnham Papers, Burnham Library, Art Institute of Chicago.

61. Neil Harris, "The Planning of the Plan, An Address Given to the 695th Regular Meeting of the Commercial Club of Chicago, 27 November 1979" (Chicago: Commercial Club, 1980); Thomas J. Schlereth, "Moody's Go-Getting Wacker Manual," *Inland Architect* 24 (April 1980): 9–11. On Burnham's sensitivity to public relations, see Burnham to Charles Moore, 11 February 1902, Burnham Papers; Charles D. Norton to Burnham, 29 October 1906, Edward H. Bennett Papers; Burnham to Norton, 3 September 1906, ibid.

62. Daniel H. Burnham, "Report on Proposed Improvements at Manila," reprinted in Moore, *Burnham*, vol. 2, 188–89.

63. James Marston Fitch, *American Building* (Boston: Houghton Mifflin, 1948), 134–35; Wilson, *City Beautiful*, 45–46; Boyer, *Urban Masses*, takes the cultural issues more seriously, as does Richard E. Foglesong, *Planning the Capitalist City: The Colonial Era to the 1920s* (Princeton: Princeton University Press, 1986), 124–66; Scott, *American City Planning*, 108; Sigfried Giedion, *Space, Time and Architecture* (Cambridge: Harvard University Press, 1941), 316.

64. Cynthia R. Field, "The City Planning of Daniel Hudson Burnham," Ph.D. diss., Department of Philosophy, Columbia University, 1974, 432.

65. Condit, *Chicago, 1910–1929*, 64; see also Tunnard, *City of Man*, 311; Scott, *American City Planning*, 109; Christopher Tunnard and Henry Hope Reed, *American Skyline* (Boston: Houghton Mifflin, 1953), 199.

66. Joan E. Draper, "Paris By The Lake: Sources of Burnham's Plan of Chicago," *Chicago Architecture, 1872–1922: Birth of a Metropolis*, ed. John Zukowsky (Munich: Prestel-Verlag, 1987), 108.

67. Arnold, *Chicago Transportation Problem*; Delano, *Chicago Railway Terminals*.

68. *Chicago Tribune,* 14 December 1853, 26 October 1867; Everett Chamberlin, *Chicago and Its Suburbs* (Chicago: T. A. Hungerford, 1874), 345; Perry R. Duis, "The Scenic Route to the Suburbs," *Chicago Magazine* 33 (March 1984): 120–22, 124; *Chicago Tribune,* 29 September 1895.

69. Montgomery Schuyler's phrase, in "A Critique of the Works of Adler and Sullivan, D. H. Burnham and Company, Henry Ives Cobb," *Architectural Record, Great American Architects Series,* no. 2 (February 1896), 3–4

70. *Chicago Tribune,* 4 July 1909.

71. Draper, "Paris by the Lake," 112.

72. *Chicago Tribune,* 11 April 1905.

73. *Chicago Tribune,* 7 April 1905, 3 March 1905.

74. Maureen A. Flanagan, *Charter Reform in Chicago* (Carbondale: Southern Illinois University Press, 1987).

75. Kathleen D. McCarthy, *Noblesse Oblige: Charity and Cultural Philanthropy in Chicago, 1849–1929* (Chicago: University of Chicago Press, 1982); Helen Lefkowitz Horowitz, *Culture and the City* (Lexington: University Press of Kentucky, 1976).

76. Douglas Sutherland, *Fifty Years on the Civic Front* (Chicago: University of Chicago Press, 1943), 9.

77. Sutherland, *Fifty Years,* 9, 27; Albion W. Small, "The Civic Federation of Chicago: A Study in Social Dynamics," *American Journal of Sociology* 1 (1895): 79–103.

78. Samuel P. Hays, "The Politics of Reform in Municipal Government in the Progressive Era," *Pacific Northwest Quarterly* 55 (October 1964): 157–69; Jon C. Teaford, *The Unheralded Triumph: City Government in America* (Baltimore: Johns Hopkins University Press, 1984).

79. *Chicago Tribune,* 11 April 1905.

80. *Chicago Tribune,* 20, 25, and 26 August 1905; see also 30 April 1905 issue.

81. *Chicago Tribune,* 26 August 1905.

82. *Cook County Court House, Report, Special Committee, New Court House, 1908* (Chicago: Hedstrom Berry, 1908); *Chicago Tribune,* 7 April 1905.

83. *Chicago Tribune,* 25 January 1905.

84. *Chicago Tribune,* 30 August 1905.

85. Edward J. Brundage to John G. Shedd, 23 April 1907, Minutes of Railway Terminal and General Plan Committees, 17 and 24 April 1907, Edward H. Bennett Papers.

86. *Inter Ocean,* 6 June and 9 August 1896. When architect Charles F. McKim saw a photograph of the Illinois Trust design, he wrote to Burnham, "It is hard to conceive of such a simple beautiful structure, with so much repose, in so ble a climate, and in the midst of such hellish surroundings! . . . It will remain a monument long after you are gone." McKim to Burnham, 15 November 1897, quoted in Moore, *Burnham,* vol. 1, 93–94.

87. Schuyler, "A Critique," 4.

Epilogue

1. Helen Lefkowitz Horowitz, *Culture and the City: Cultural Philanthropy in Chicago from the 1880s to 1917* (Lexington: University Press of Kentucky, 1976); Carl S. Smith, *Chicago and the American Literary Imagination, 1880–1920* (Chicago: University of Kentucky Press, 1984).

2. John Munn Journal, 30 April 1849, Manuscript Division, Chicago Historical Society.

3. Fredrika Bremer, *The Homes of the New World,* 2 vols. (New York: Harper & Bros., 1853), quoted in Bessie Louise Pierce, *As Others See Chicago* (Chicago: University of Chicago Press, 1933), 129–30.

4. Karl Baedecker, ed., *The United States, with an Excursion into Mexico* (New York: Charles Scribner's Sons, 1899), 311.

5. George Wharton James, *Chicago's Dark Places: Investigations by a Corps of Specially Appointed Commissioners, Edited and Arranged by the Chief Commissioner* (Chicago: Craig Press, 1891), 13–15.

6. Ibid; Ross Miller's *American Apocalypse: The Great Fire and the Myth of Chicago* (Chicago: University of Chicago Press, 1990), argues that the World's Columbian Exposition was built to "distract" visitors from the "city's real life," pp. 195–97. I would argue that although such a strategy of disguise was at work at the fair, it had also profoundly influenced the building and structuring of central elements of the real city in the course of the nineteenth century.

7. Mark Girouard, *Cities and People: A Social and Architectural History* (New Haven: Yale University Press, 1985), 318–19.

8. John J. Flinn, *The Standard Guide of Chicago* (Chicago: Standard Guide Company, 1893), 99–103.

9. William S. McCormick to Cyrus Hall McCormick, 28 December 1862, McCormick Papers, Manuscript Division, Wisconsin Historical Society Library, Madison.

10. James Fullarton Muirhead, *America, the Land of Contrasts* (New York: John Lane Company, 1898), 207–8; Harvey W. Zorbaugh, *Gold Coast and Slum: A Sociological Study of Chicago's Near North Side* (Chicago: University of Chicago Press, 1929).

11. Louis Sullivan, "The Tall Office Building Artistically Considered," *Lippincott's Magazine* 57 (March 1896): 403–9; Karen Halttunen, *Confidence Men and Painted Women: A Study of Middle-Class Culture in America, 1830–1870* (New Haven: Yale University Press, 1982), examines middle-class Americans' obsession with "sincerity."

12. Mel Scott, *American City Planning Since 1890* (Berkeley: University of California Press, 1969); Richard E. Fogelsong, *Planning the Capitalist City: The Colonial Era to the 1920s* (Princeton: Princeton University Press, 1986).

Bibliography

Ahlstrom, Sydney E. *A Religious History of the American People*. New Haven: Yale University Press, 1972.

Andreas, Alfred T. *History of Chicago, from the Earliest Period to the Present Time*. 3 vols. Chicago: A. T. Andreas, 1884–86.

Arnold, Matthew. *Culture and Anarchy: An Essay in Political and Social Criticism*. London: Smith, Elder, 1869.

Avrich, Paul. *The Haymarket Tragedy*. Princeton: Princeton University Press, 1984.

Barth, Gunther. *City People: The Rise of Modern City Culture in Nineteenth Century America*. New York: Oxford University Press, 1980.

Bellamy, Edward. *Looking Backward, 2000–1887*. Boston: Ticknor, 1888.

Bender, Thomas. *Toward an Urban Vision: Ideas and Institutions in Nineteenth-Century America*. Louisville: University of Kentucky Press, 1975.

Bennett, Edward H. Papers. Burnham Library, Art Institute of Chicago, Chicago.

Beveridge, Charles E. "Frederick Law Olmsted's Theory of Landscape Design." *Nineteenth Century* 3 (summer 1977): 38–45.

Biographical Sketches of the Leading Men of Chicago, Written by the Best Talent in the Northwest. Chicago: Wilson & St. Clair, 1868.

Bluestone, Daniel M. "Detroit's City Beautiful and the Problem of Commerce." *Journal of the Society of Architectural Historians* 47 (September 1988): 245–62.

Bluestone, Daniel M. "From Promenade to Park: The Gregarious Origins of Brooklyn's Park Movement." *American Quarterly* 39 (winter 1987): 529–50.

Boyer, M. Christine. *Dreaming the Rational City: The Myth of American City Planning*. Cambridge: MIT Press, 1983.

Boyer, Paul. *Urban Masses and Moral Order in America, 1820–1920*. Cambridge: Harvard University Press, 1978.

Burnham, Daniel H. Papers. Burnham Library, Art Institute of Chicago.

Burnham, Daniel H., and Edward H. Bennett. *Plan of Chicago*. Chicago: Commercial Club, 1909.

Bushman, Richard L. "American High-Style and Vernacular Cultures." In *Colonial British America: Essays in the New History of the Early Modern Era*, ed. Jack P. Greene and J. R. Pole. Baltimore: Johns Hopkins University Press, 1984.

Chamberlin, Everett. *Chicago and Its Suburbs*. Chicago: T. A. Hungerford, 1874.

Chandler, Alfred D., Jr. *The Visible Hand: The Managerial Revolution in American Business*. Cambridge: Harvard University Press, 1977.

Chicago Board of Trade. Papers. Manuscript Archives, University of Illinois, Chicago Circle.

Cleveland, Horace William Shaler. *The Public Grounds of Chicago: How to Give Them Character and Expression.* Chicago: Charles D. Lakey, 1869.

Cmiel, Kenneth. *Democratic Eloquence: The Fight over Popular Speech in Nineteenth-Century America.* New York: William Morrow, 1990.

Condit, Carl. *The Chicago School of Architecture: A History of Commercial and Public Building in the City Area, 1875–1925.* Chicago: University of Chicago Press, 1964.

Condit, Carl. *The Rise of the Skyscraper.* Chicago: University of Chicago Press, 1952.

Cranz, Galen. *The Politics of Park Design: A History of Urban Parks in America.* Cambridge: MIT Press, 1982.

Ciucci, Giorgio. Francesco Dal Co, Mario Manieri-Elia, and Manfredo Tafuri. *The American City: From the Civil War to the New Deal.* 1973. Trans. Barbara Luigia La Penta. Cambridge: MIT Press, 1979.

Douglas, Ann. *The Feminization of American Culture.* New York: Knopf, 1977.

Downing, Andrew Jackson. *Rural Essays.* New York: G. P. Putnam, 1853.

Draper, Joan E. *Edward H. Bennett: Architect and City Planner, 1874–1954.* Chicago: Art Institute of Chicago, 1982.

Draper, Joan E. "Paris by the Lake: Sources of Burnham's Plan of Chicago." In *Chicago Architecture, 1872–1922: Birth of A Metropolis,* ed. John Zukowsky. Munich: Prestel-Verlag, 1987.

Duis, Perry R. *Chicago Creating New Traditions.* Chicago: Chicago Historical Society, 1976.

Duis, Perry R. *The Saloon: Public Drinking in Chicago and Boston, 1880–1920.* Urbana: University of Illinois Press, 1983.

Duncan, Hugh D. *Culture and Democracy: The Struggle for Form in Society and Architecture in Chicago and the Middle West during the Life and Times of Louis H. Sullivan.* Totowa, N.J.: Bedminster Press, 1965.

Flinn, John J. *Chicago, the Marvelous City of the West: A History, An Encyclopedia, and a Guide.* Chicago: The Standard Guide Company, 1892.

Flinn, John J. *The Standard Guide of Chicago.* Chicago: The Standard Guide Company, 1893.

Foglesong, Richard E. *Planning the Capitalist City: The Colonial Era to the 1920s.* Princeton: Princeton University Press, 1986.

Giedion, Sigfried. *Space, Time and Architecture: The Growth of a New Tradition.* Cambridge: Harvard University Press, 1941.

Gilbert, Paul T., and Charles L. Bryson. *Chicago and Its Makers.* Chicago: F. Mendelsohn, 1929.

Girouard, Mark. *Cities and People: A Social and Architectural History.* New Haven: Yale University Press, 1985.

Halttunen, Karen. *Confidence Men and Painted Ladies: A Study of Middle-Class Culture in America, 1830–1870.* New Haven: Yale University Press, 1982.

Harris, Neil. *The Artist in American Society: The Formative Years, 1790–1860.* New York: George Braziller, 1966.

Harris, Neil. *Cultural Excursions: Marketing Appetites and Cultural Tastes in Modern America.* Chicago: University of Chicago Press, 1990.

Hines, Thomas S. *Burnham of Chicago: Architect and Planner.* New York: Oxford University Press, 1974.

Hobsbawn, Eric J. *The Age of Empire, 1875–1914.* New York: Pantheon Books, 1987.

Hoffmann, Donald. *The Architecture of John Wellborn Root.* Baltimore: Johns Hopkins University Press, 1973.

Holt, Glen E. "Private Plans for Public Squares: The Origins of Chicago's Park System, 1850–1875." *Chicago History* 8 (fall 1979): 173–84.

Horowitz, Helen Lefkowitz. *Culture and the City: Cultural Philanthropy in Chicago from the 1880s to 1917.* Lexington: University Press of Kentucky, 1976.

Hutchison, William R. "Disapproval of Chicago: The Symbolic Trial of David Swing." *Journal of American History* 59 (June 1972): 30–47.

Industrial Chicago. Vols. 1–2, *The Building Interests.* Chicago: Goodspeed Publishing, 1891.

Jackson, John Brinckerhoff. *American Space: The Centennial Years, 1865–1876.* New York: W. W. Norton, 1972.

Jaher, Frederic Cople. *The Urban Establishment: Upper Strata in Boston, New York, Charleston, Chicago, and Los Angeles.* Urbana: University of Illinois Press, 1982.

Jenney, William Le Baron, and Sanford E. Loring. *Principles and Practice of Architecture, comprising Forty-Six Folio Plates of Plans, Elevations and Details of Churches, Dwellings and Stores Constructed By the Authors. Also An Explanation and Illustrations of the French System of Apartment Houses, And Dwellings For the Laboring Classes, Together with Copious Text.* Chicago: Cobb, Pritchard, 1869.

Jevne & Almini. *Chicago Illustrated.* Chicago, 1867–68.

Jordy, William H. *American Buildings and Their Architects: Progressive and Academic Ideals at the Turn of the Twentieth Century.* Garden City, N.Y.: Doubleday, 1972.

Jordy, William H. "The Commercial Style and the 'Chicago School.'" Review of *The Chicago School of Architecture,* by Carl Condit, in *Perspectives in American History* 1 (1967): 390–400.

Kasson, John F. *Civilizing the Machine: Technology and Republican Values in America, 1776–1900*. New York: Grossman, 1976.

Kowsky, Francis R. "Municipal Parks and City Planning: Frederick Law Olmsted's Buffalo Park and Parkway System." *Journal of the Society of Architectural Historians* 46 (March 1987): 49–64.

Landau, Sarah Bradford. "The Tall Office Building Artistically Reconsidered: Arcaded Buildings in New York, c. 1850–1890." In *In Search of Modern Architecture: A Tribute to Henry-Russell Hitchcock*, ed. Helen Searing. Cambridge: MIT Press, 1982.

Lane, George. *Chicago Churches and Synagogues*. Chicago: Loyola University Press, 1981.

Larson, Gerald R., and Roula Mouroudellis Geraniotis, "Toward a Better Understanding of the Evolution of the Iron Skeleton Frame in Chicago." *Journal of the Society of Architectural Historians* 46 (March 1987): 39–48.

Lears, T. J. Jackson. *No Place of Grace: Anti-Modernism and the Transformation of American Culture, 1880–1920*. New York: Pantheon Books, 1981.

Leeuwen, Thomas A. P. van. *The Skyward Trend of Thought: The Metaphysics of the American Skyscraper*. Cambridge: MIT Press, 1988.

Lewis, Russell. "Everything under One Roof: World's Fairs and Department Stores in Paris and Chicago." *Chicago History* 12 (fall 1983): 28–47.

Mayer, Harold M., and Richard C. Wade. *Chicago: Growth of a Metropolis*. Chicago: University of Chicago Press, 1969.

McCormick, Cyrus Hall. Papers. Wisconsin Historical Society, Madison.

Miller, Ross. *American Apocalypse: The Great Fire and the Myth of Chicago*. University of Chicago Press, 1990.

Monroe, Harriet. *John Wellborn Root*. New York: Houghton, Mifflin, 1896.

Mumford, Lewis. *Sticks and Stones: A Study of American Architecture and Civilization*. New York: Horace Liveright, 1924.

Newton, Joseph Fort. *David Swing: Poet Preacher*. Chicago: Unity Publishing Company, 1909.

Ogden, William B. Papers. Manuscript Division, Chicago Historical Society.

Olmsted, Frederick Law. Papers. Manuscript Division, Library of Congress, Washington, D.C.

Olmsted, Vaux and Company. *Report accompanying Plan for Laying Out the South Park, March 1871*. Chicago: South Park Commission, 1871.

Olsen, Donald J. *The City as a Work of Art: London, Paris, Vienna*. New Haven: Yale University Press, 1986.

Olsen, Donald J. *The Growth of Victorian London*. Hammondsworth: Peregrine, 1979.

The Parks and Property Interests of the City of Chicago. Chicago: Western News, 1869.

Peterson, Jon A. "The City Beautiful Movement: Forgotten Origins and Lost Meanings." *Journal of Urban History* 2 (August 1976): 415–34.

Phillips, George S. *Chicago and Her Churches*. Chicago: Myers and Chandler, 1868.

Philpott, Thomas Lee. *The Slum and the Ghetto: Neighborhood Deterioration and Middle-Class Reform, Chicago, 1880–1930*. New York: Oxford University Press, 1978.

Pierce, Bessie L. *A History of Chicago*. 3 vols. New York: Knopf, 1937–56.

Public Buildings Service. Records. General Correspondence, record group 121. National Archives, Washington, D.C.

Rand McNally. *Bird's-Eye View and Guide to Chicago*. Chicago, 1893.

Ranney, Victoria Post. *Olmsted in Chicago*. Chicago: R. R. Donnelley and Sons, 1972.

Rauch, John H. *Public Parks: Their Effects upon the Moral, Physical and Sanitary Conditions of the Inhabitants of Large Cities with Special Reference to the City of Chicago*. Chicago: S. C. Griggs, 1869.

Randall, Frank A. *History of the Development of Building Construction in Chicago*. Urbana: University of Illinois Press, 1949.

Reps, John W. *The Making of Urban America: A History of City Planning in the United States*. Princeton: Princeton University Press, 1965.

Reps, John W. *Town Planning in Frontier America*. Princeton: Princeton University Press, 1965.

Robinson, Charles Mulford. *Modern Civic Art or the City Made Beautiful*. New York: G. P. Putnam's Sons, 1903.

Root, John W. "A Great Architectural Problem." *Inland Architect* 15 (June 1890): 67–71.

Rosenzweig, Roy. *Eight Hours for What We Will: Workers and Leisure in an Industrial City, 1870–1920*. Cambridge: Cambridge University Press, 1983.

Roth, Leland M. *McKim, Mead and White, Architects*. New York: Harper and Row, 1983.

Ryan, Mary P. *Cradle of the Middle Class: The Family in Oneida County, N.Y., 1790–1865*. Cambridge: Cambridge University Press, 1985.

Ryan, Mary P. *Women in Public: Between Banners and Ballots, 1825–1880*. Baltimore: Johns Hopkins University Press, 1990.

Sandburg, Carl. *Chicago Poems*. New York: Henry Holt, 1916.

Schmitt, Peter. *Back to Nature: The Arcadian Myth in Urban America*. New York: Oxford University Press, 1969.

Schorske, Carl E. *Fin-De Siecle Vienna, Politics and Culture*. New York: Knopf, 1980.

Schultz, Stanley K. *Constructing Urban Culture: American Cities and City Planning, 1800–1920*. Philadelphia: Temple University Press, 1989.

Schuyler, David. *The New Urban Landscape: The Redefinition of City Form in Nineteenth-Century America*. Baltimore: Johns Hopkins University Press, 1986.

Scott, Mel., Jr. *American City Planning since 1890*. Berkeley: University of California Press, 1969.

Simon, Andreas. *Chicago, the Garden City: Its Magnificent Parks, Boulevards and Cemeteries*. Chicago: Franz Gindele, 1895.

Siry, Joseph. *Carson Pirie Scott: Louis Sullivan and the Chicago Department Store*. Chicago: University of Chicago Press, 1988.

Smith, Carl S. *Chicago and the American Literary Imagination, 1880–1920*. Chicago: University of Chicago Press, 1984.

Stansell, Christine. *City of Women: Sex and Class in New York, 1789–1860*. New York: Knopf, 1986.

Stead, William T. *If Christ Came to Chicago: A Plea for the Union of All Who Love in the Service of All Who Suffer*. London: Review of Reviews, 1894.

Stern, Robert A. M., Gregory Gilmartin, and John Montague Massengale, *New York 1900 Metropolitan Architecture and Urbanism, 1890–1915*. New York: Rizzoli, 1983.

Sullivan, Louis. "The Tall Office Building Artistically Considered." *Lippincott's Magazine* 57 (March 1896): 403–9.

Trachtenberg, Alan. *The Incorporation of America: Culture and Society in the Gilded Age*. New York: Hill & Wang, 1982.

Turak, Theodore. *William Le Baron Jenney: A Pioneer of Modern Architecture*. Ann Arbor: UMI Research Publications, 1986.

Upton, Dell and John M. Vlach, eds. *Common Places: Readings in American Vernacular Architecture*. Athens: University of Georgia Press, 1986.

Upton, Dell. *Holy Things and Profane: Anglican Parish Churches in Colonial Virginia*. Cambridge: MIT Press, 1986.

Veblen, Thorstein. *The Theory of the Leisure Class: A Study of Economic Institutions*. New York: Macmillan, 1899.

Warner, Sam Bass, Jr. *Streetcar Suburbs: The Process of Growth in Boston, 1870–1900*. Cambridge: Harvard University Press, 1962.

Welter, Barbara. "The Cult of True Womanhood, 1820–1860." *American Quarterly* 18 (summer 1966): 131–75.

Williams, Raymond. *The Country and the City*. New York: Oxford University Press, 1973.

Wille, Lois. *Forever Open, Clear and Free: The Struggle for Chicago's Lakefront*. Chicago: Henry Regnery, 1972.

Wilson, William H. *The City Beautiful Movement*. Baltimore: Johns Hopkins University Press, 1989.

Wit, Wim de, ed. *Louis Sullivan: The Function of Ornament*. New York: W. W. Norton, 1986.

Wright, Gwendolyn. *Moralism and the Model Home: Domestic Architecture and Cultural Conflict in Chicago, 1873–1913*. Chicago: University of Chicago Press, 1980.

Wright, John S. *Chicago Past, Present, Future: Relations to the Great Interior and to the Continent*. Chicago: Horton & Leonard, 1868.

Zaitzevsky, Cynthia. *Frederick Law Olmsted and the Boston Park System*. Cambridge: Harvard University Press, 1982.

Zukowsky, John, ed. *Chicago Architecture, 1872–1922: Birth of A Metropolis*. Munich: Prestel-Verlag, 1987.

Zunz, Olivier. *Making America Corporate, 1870–1920*. Chicago: University of Chicago Press, 1990.

Index

All italicized page numbers refer to pages where illustrations appear separately from indexed text.

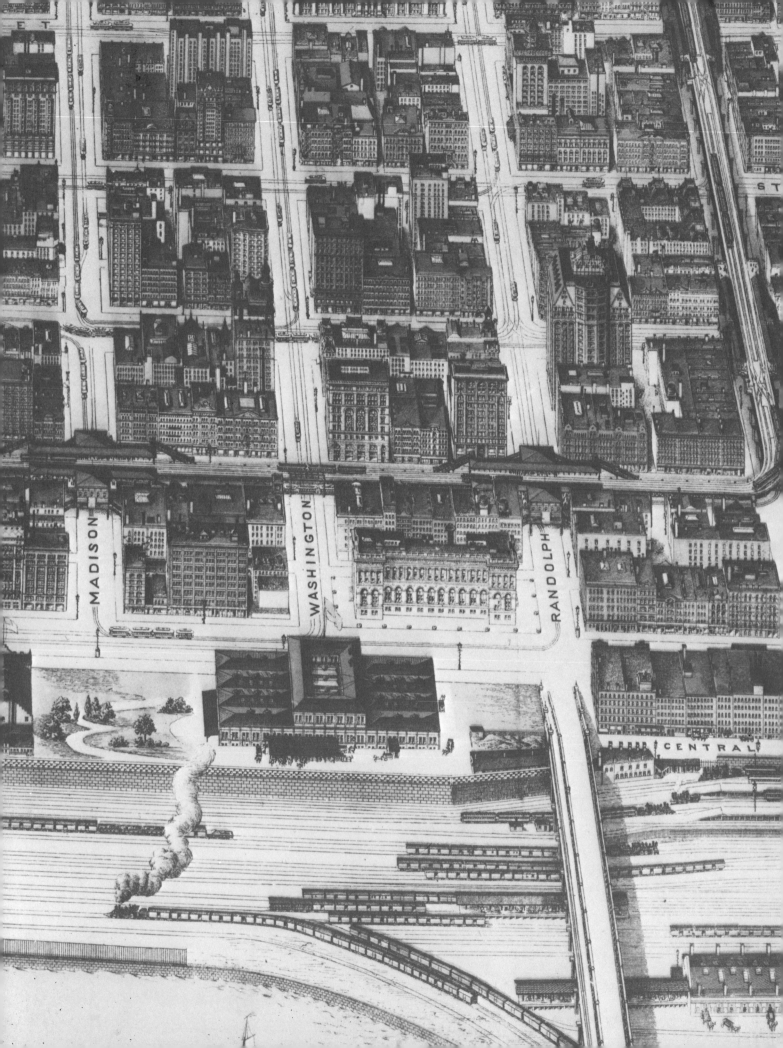